£0.[illegible]

Th

About the Book

The third edition of
designed to give the
the main tanks and o
with the world's armies of today.

Vehicles are grouped under country of origin, and subdivided into main battle tanks, reconnaissance vehicles, self-propelled artillery, mechanized infantry combat vehicles and armoured personnel carriers, anti-aircraft weapon carriers and tank destroyers. A brief history of each vehicle is given, together with the countries which use it and technical data (crew, dimensions, speed, range, weight, armament – including types of ammunition used – and engine details). Special characteristics of particular vehicles, details of variants and manufacturers are also provided.

The photographs and specially prepared silhouettes will enable the reader to identify a vehicle when he sees it in the flesh, in photographs or on television or film, and the index at the back will help him quickly find mention of it in the book. The Introduction gives a comprehensive overview of current developments in the world's armoured fighting vehicles.

About the Author

Charles Messenger is a professional military historian and defence commentator. He was for twenty-one years an officer in the Royal Tank Regiment, and still maintains active links with the Army. His military career took him to the Near East, Germany and the USA, and he has had personal experience of many of the vehicles detailed in this book, both with his regiment and as a staff officer. A graduate of the Army Staff College and an MA in Modern History from Oxford, he has written several books on twentieth-century warfare – air, land and sea – as well as on contemporary defence matters. He also contributes articles and book reviews to a wide range of defence and historical journals, as well as lecturing on military history and defence matters.

The *Observer's* series was launched in 1937 with the publication of *The Observer's Book of Birds*. Today, fifty years later, paperback *Observers* continue to offer practical, useful information on a wide range of subjects, and with every book regularly revised by experts, the facts are right up-to-date. Students, amateur enthusiasts and professional organizations alike will find the latest *Observers* invaluable.

'Thick and glossy, briskly informative' – *The Guardian*

'If you are a serious spotter of any of the things the series deals with, the books must be indispensable' – *The Times Educational Supplement*

OBSERVERS

TANKS

AND OTHER ARMOURED VEHICLES

Charles Messenger

With silhouettes by Michael Badrocke

FREDERICK WARNE

FREDERICK WARNE
Penguin Books Ltd, Harmondsworth, Middlesex, England
Viking Penguin Inc., 40 West 23rd Street, New York, New York 10010, U.S.A.
Penguin Books Australia Ltd, Ringwood, Victoria, Australia
Penguin Books Canada Limited, 2801 John Street, Markham, Ontario, Canada L3R 1B4
Penguin Books (N.Z.) Ltd, 182–190 Wairau Road, Auckland 10, New Zealand

First published in hardback, 1981
First published in paperback, 1984
Third edition 1987

Photographic Acknowledgements

The author and publishers would like to thank the following for photographs published in this book on the page numbers indicated: Steyr-Daimler-Puch, *12, 14;* Engesa SA, *18, 20;* EC des Armées, *26, 30;* Panhard et Levassor, *32, 40;* Renault, *34;* Keystone Press Agency, *38;* Creusot-Loire, *42;* ATS Roanne, *44, 46;* Krauss Maffei AG, *54;* Thyssen Henschel, *56, 62;* Israeli Embassy, London, *70;* Ramta Structures and Systems, *72;* Soltam Co Ltd, *74;* OTO-Melara, *76, 80;* Mitsubishi Heavy Industries, *82, 84;* DAF, *88;* BRAVIA, *90;* Empresa Nacional, *94;* AB Bofors, *96, 100;* Swiss Embassy, London, *102;* General Motors of Canada Ltd, *104;* Ministry of Defence (UK), *108, 110;* Vickers Ltd, *112;* GKN Sankey, *126, 128;* Chrysler Corporation, *136;* General Dynamics Land Systems, *138;* Dutch Ministry of Defence, *140;* FMC Corporation, *152, 156.*

ISBN 0–7232–3511–2

Printed and bound in Great Britain by
Butler & Tanner Ltd
Frome and London

INTRODUCTION

As I remarked in the Introduction to the previous edition, (1984), the number of different types of armoured fighting vehicle (AFV) and, indeed, the number of countries manufacturing them, continue to increase. Limited space means that not every AFV can be covered, so in general I have restricted myself to the main representative vehicles which are in or about to enter service; and I have concentrated on *fighting* rather than support vehicles, and on those which the observer is most likely to see. As in the previous edition, I have also included examples from some of the smaller AFV producing countries. In one or two cases I have broken this rule, but this has been to reflect new trends. Thus, the FRG Weisel will not enter service for another 2–3 years, but does emphasize trends towards ultra-light vehicles.

General developments

Since 1984 there have been no dramatic changes, such as new main battle tanks being introduced. Apart from the Iran–Iraq war and the fighting in Afghanistan, which continue to rumble on, there have been no significant combat tests of AFVs. Nevertheless, two interesting trends can be identified. The first, already referred to, is the interest in very light armoured vehicles. There are two reasons for this. The first is strategic and concerns the growth of intervention forces such as the US Central Command (what used to be known as the Rapid Deployment Force) and the French Force Action Rapide. These rely on airlift, of which there never appears to be enough, and the forces involved have of necessity to be on light scales, which means that supporting firepower has to be sacrificed. Yet, there is the very real fear that they will find themselves deployed to a country where the enemy possesses numbers of AFVs and they need weapons which can deal with this threat. Likewise, at the lower operational and tactical levels, there is renewed interest in the use of airmobile forces (those which are deployed to battle by helicopter), especially as quick reaction elements which can be deployed to block surprise enemy armoured thrusts or be sent ahead of attacking forces to seize vital objectives such as bridges and defiles. Again, there is a requirement that they should have mobile anti-tank weapons and hence the introduction of the FRG Weisel and French M11 VBL, which can be carried by helicopter and mount ATGW systems.

The second significant development concerns the traditional gun-armour race. Most main battle tanks are now, or shortly will be, protected by what is popularly known as Chobham armour after the British Military Vehicles and Engineering Establishment at Chobham, Surrey. This is very effective against the current generation of HEAT warheads with which ATGW systems are equipped. The only way in which these warheads can be improved is to increase their diameter and this is what is happening. The penalty to be paid, however, is that in many cases the turrets in which these ATGW systems are fitted are too small to accommodate the improved warheads and hence a certain amount of re-engineering is now taking place.

Much discussion continues on the future of the main battle tank as it exists today in its 50–60 metric ton form. The argument that its days are numbered because of the appearance of 'top attack' smart munitions and sub-munitions persists as strongly as ever. Many experts believe that the solution to the dilemma is the 'small' tank in the 20–30 metric ton range, probably with an externally mounted gun and the crew all in the hull rather than in a turret. The main objection to this is that with no direct vision devices tank commanders would lose their ability to 'sense' the battle. Nevertheless, experimentation with this concept continues in many countries, although only the Swedes appear to be committed to it for the replacement of their S-Tank.

Soviet tank designations

Confusion persists in Western circles over the correct designation of the Soviet T-64 and T-72. Both have undergone a number of modifications in recent years, of which the most recent appears to have been the installation of an ATGW system in both. The Americans have classified these versions as T-64B and T-80, but other countries believe that the latter is T-74 and that T-80 is still under development. This reflects the perennial problem of trying to establish accurate data on Soviet equipment, which says much for the continuing effectiveness of Soviet security.

New Israeli armour

In the past few years the Israelis have developed a new armour called BLAZER. This consists of metal boxes filled with explosive which are attached to the outside of the AFV. When struck by a chemical energy warhead (HEAT, HESH), the explosive is detonated, producing shock waves which radiate outwards, thus helping to neutralize the effect of the warhead. All Israeli tanks now have this form of reactive armour, especially around the turret, and it is likely that other nations will also follow their example.

Third World countries and AFVs

At a time when even the most advanced of Western industrial nations are finding it increasingly difficult to find the money to be able to fund AFV development and production programmes to the extent that they require, the problem facing the poorer Third World countries is even more severe. At root lies the ever increasing technological sophistication of AFVs, especially fire control systems and night vision devices. This means that few countries can afford to maintain a fleet of modern AFVs by buying 'off the shelf'. There are two solutions. The first is to manufacture under licence, which an increasing number of countries are doing. This, however, requires a reasonable industrial base in the first place, which many of the poorer and more technically backward states do not possess. The other, which has been recognized by many AFV producing companies, is to upgrade obsolescent AFVs through the provision of modernization kits. An example of this is the US M48 main battle tank.

Replacing its 90 mm gun with a 105 mm model and providing an integrated fire control system, including laser rangefinder and computer, radically improves the effectiveness of a tank which is over thirty years old, and a number of countries which use it are already installing these kits. Another example is the mounting of British 105 mm guns in Soviet T-54/55s.

Collaborative projects

The sharply spiralling costs in developing modern weapon systems can be reduced through international collaboration. This is especially recognized in Western Europe, where national defence industries have for a long time faced fierce competition from their US counterparts, who usually enjoy the major advantage of longer production runs and hence lower unit cost prices. European collaborative efforts on combat aircraft and helicopters have been common for some years, but for AFVs progress in this field has not been so dramatic. There were attempts in the 1970s to set up an Anglo-German project for a new main battle tank, but this foundered on an inability to resolve conflicting design priorities, especially as to whether protection should be put ahead of mobility, a policy traditionally favoured by the British. There is also a Franco–German study for an MBT of the 1990s, but since the French are designing their own tank, to be called the General Leclerc, it is unlikely to come to anything. One other major project which has run for some years has been the SP-70 self-propelled artillery howitzer. The team developing this was made up of the FRG, Italy and UK but constant problems, both technical and funding, resulted in it failing to meet laid-down specifications and at the end of 1986 it was cancelled. Whether this failure will discourage further European collaboration on major AFV projects remains to be seen.

Individual vehicle entries

The reader should be aware of what is meant by various terms used in the 'Vehicle data' section. Weight is combat weight, that is the vehicle, crew, ammunition, full fuel tanks and all the ancillary equipment required to operate the AFV in battle.

Maximum speed and range are for the vehicle on roads, with range being that for a full tank of fuel and should be reduced by some 40% for cross-country.

ABBREVIATIONS USED IN THE TEXT

AA	Anti-aircraft
ACV	Armoured Command Vehicle
AEV	Armoured Engineer Vehicle
AFV	Armoured Fighting Vehicle
AIFV	Armoured Infantry Fighting Vehicle
APC	Armoured Personnel Carrier
APDS	Armour-Piercing Discarding Sabot
APDSFS	Armour-Piercing Discarding Sabot Fin-Stabilized
APHE	Armour-Piercing High Explosive
ARRV	Armoured Recovery and Repair Vehicle
ARV	Armoured Recovery Vehicle
AT	Anti-Tank
ATGW	Anti-Tank Guided Weapon
AVLB	Armoured Vehicle Launched Bridge
AVRE	Armoured Vehicle Royal Engineers
BARV	Beach Armoured Recovery Vehicle
CEV	Combat Engineer Vehicle
CFV	Cavalry Fighting Vehicle
DDR	German Democratic Republic (East)
FRG	Federal Republic of Germany (West)
FVSV	Fighting Vehicle Systems Carrier
HE	High Explosive
HEAT	High Explosive Anti-Tank
HEATFS	High Explosive Anti-Tank Fin-Stabilized
HEP	High Explosive Piercing
HESH	High Explosive Squash Head
HVAP	High Velocity Armour-Piercing
IFV	Infantry Fighting Vehicle
IFCS	Integrated Fire Control System
LLTV	Low Light Television

MBT	Main Battle Tank	OP	Observation Post
mg	machine gun	rmg	ranging machine gun
MICV	Mechanized Infantry Combat Vehicle	SAM	Surface-to-Air Missile
MLRS	Multiple Launch Rocket System	shp	shaft horse-power
NBC	Nuclear, Biological and Chemical	SSM	Surface-to-Surface Missile

CONVERSION FROM METRIC UNITS

Length	1 centimetre (cm)	= 0.39 inches
	1 metre (m)	= 39.37 inches
	1 kilometre (km)	= 0.62 miles (or 5 miles = 8 km)
Weight	1 kilogram (kg)	= 2.21 lbs
	1000 kg	= 1 metric tonne = 0.984 tons
Calibre	7.62 mm	= 0.30 inches
	12.7 mm	= 0.50 inches
	76 mm	= 17 pounder (pdr)
	84 mm	= 20 pounder

DIMENSIONS

The principal dimensions in the text are as precise as possible, given the wide variety of sources. In consequence there are inconsistencies in the number of decimal places; these have been retained in the interests of accuracy.

Role Mechanized infantry combat vehicle.

History Origins lie in the development by Saurer of an APC, the 4K 4F, in the late 1950s. This entered service with the Austrian Army in 1961 and production was completed in 1969. First prototype of 4K 7FA appeared in 1976, and it entered Austrian Army service a year later. The engine and transmission are as in Jagdpanzer SK 105 tank destroyer.

Employment Austria, Bolivia, Greece (known as Leonidas), Nigeria.

Vehicle data *Crew* 2 (driver, gunner) + 8 infantrymen, *height* 1.60 m (excluding turret), *width* 2.50 m, *length* 5.87 m, *weight* 14,800 kg, *max. speed* 65.5 kph, *range* 520 km, *armament* 1 × 12.7 mm, 1 × 7.62 mm pintle-mounted mg on rear deck, *engine* Steyr 7FA 6-cylinder diesel developing 320 bhp at 2300 rpm.

Special characteristics Passive night vision equipment and NBC protection system. Four ball mounts are fitted in the hull (two on each side) for infantry to engage with mgs from inside vehicle.

Variants 4K 7FA-Fü, command vehicle, 4K 7FA/SAN ambulance. 4K 7FA-KvPz 1/90 fire support combat vehicle (90 mm gun), 4K 7FA-SPz/300 MICV (30 mm Rarden gun), 4K 7FA GrW 81 (81 mm mortar). Prototypes with 2 × 20 mm and 2 × 30 mm AA guns also exist.

Manufacturer Steyr-Daimler-Puch.

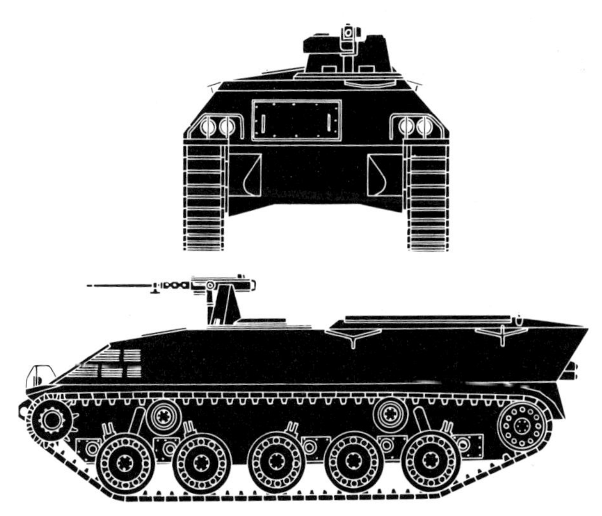

JAGDPANZER SK 105 KÜRASSIER

Role Tank destroyer.

History To meet Austrian Army requirement for a tank destroyer, Saurer (taken over by Steyr-Daimler-Puch in 1970) began developing the SK 105 in 1965. Using many components of the 4K 4F APC, the first prototype appeared two years later, followed by five pre-production vehicles in 1971. It entered service with the Austrian Army soon afterwards. The turret is an improved version of the FL-12 found on AMX-13 (see separate entry).

Employment Argentina, Austria, Bolivia, Morocco, Nigeria, Tunisia.

Vehicle data *Crew* 3 (commander, gunner, driver), *height* 2.53 m, *width* 2.50 m, *length* 5.58 m (7.76 m with gun fully forward), *weight* 17,500 kg, *max. speed* 65.5 kph, *range* 520 km, *armament* 1 × 105 mm rifled gun (fires

APFSDS, HEAT, HE, Smoke), 1 × 7.62 mm coaxial mg, *engine* Steyr 7FA 6-cylinder diesel developing 320 bhp at 2300 rpm (without cooling air-blower operating).

Special characteristics Semi-automatic loader (two six-pound revolver-type magazines), laser rangefinder, active and passive night-viewing devices, NBC protection system.

Variants Greif ARV with crane and dozer blade, 4KH 7FA-PL engineer tank similar to Greif, but large dozer blade and no winch, 4KH 7FA-FA driver training version with enclosed cabin instead of turret. 105/AZ with fully stabilised oscillating turret, computerised fire control system, fully automatic loader, 4KH 7PA-Pi engineer vehicle with winch, dozer blade and excavator.

Manufacturer Steyr-Daimler-Puch.

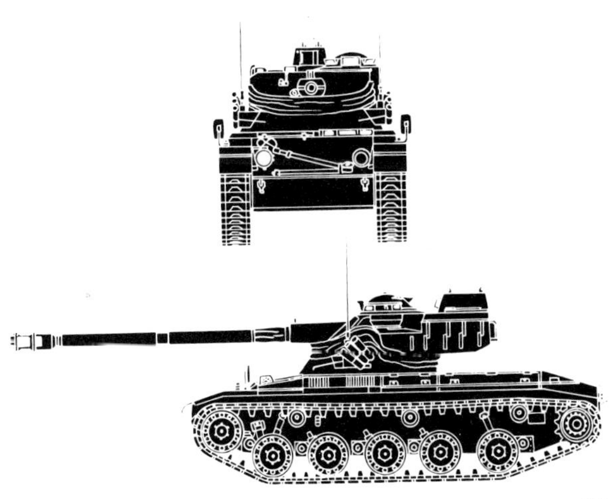

Role Main battle tank.

History Developed by Engesa to meet both a Brazilian Army requirement and a potential export market, the first prototype appeared in autumn 1984. The UK firm of Vickers designed and produced two prototype turrets (105 mm, 120 mm) and production started in 1986.

Employment Brazil.

Vehicle data *Crew* 4 (commander, gunner, driver, loader), *height* 2.371 m, *width* 3.26 m, *length* 7.08 m, *weight* 39,000 kg, *max. speed* 70 kph, *range* 550 km, *armament* 1 × 105 mm (rifled), 1 × 7.62 mm coaxial mg,

1 × 7.62 mm AA mg, *engine* diesel developing 1000 bhp at 2300 rpm.

Special characteristics Laser rangefinder, night fighting devices, NBC protection available.

Variants Turret mounting 120 mm smoothbore gun; 12.7 mm AA mg can be fitted.

Manufacturer Engesa, São Paulo.

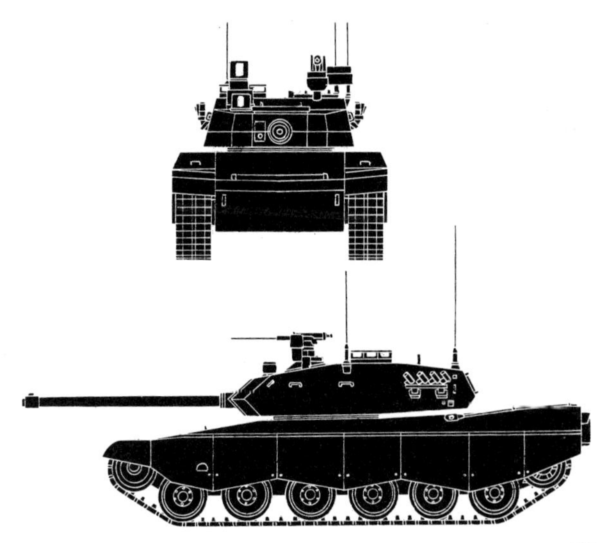

Role Armoured car.

History Development began in 1970, with the first prototype being produced that same year. It incorporates many of the components used in the Urutu APC. Entered Brazilian Army Service in 1975.

Employment Bolivia, Brazil, Chad, Chile, Colombia, Cyprus, Gabon, Iraq, Libya, Tunisia, Uruguay, Zimbabwe.

Vehicle data (Mk 3) *Crew* 3 (commander, gunner, driver), *height* 2.33 m, *width* 2.44 m, *length* 5.18 m (5.99 m with gun fully forward), *weight* 10,750 kg, *max. speed* 100 kph, *range* 880 km, *armament* 1 × Cockerill low-pressure 90 mm (HEAT, HESH), 1 × 7.62 mm coaxial mg, *engine* Mercedes-Benz (Brazil) OM-352-A diesel developing 172 bhp at 2800 rpm.

Special characteristics Night-fighting and -driving aids, laser rangefinder, amphibious with special flotation kit.

Variants Mk 1 pre-production model with 37 mm gun; Mk 2 French H-90 turret with French GIAT 90 mm (see photograph); Mk 3 with Engesa EC-90 gun; Mk 4 General Motors Detroit diesel; Mk 5 with Mercedes-Benz diesel.

Manufacturer Engesa, São Paulo.

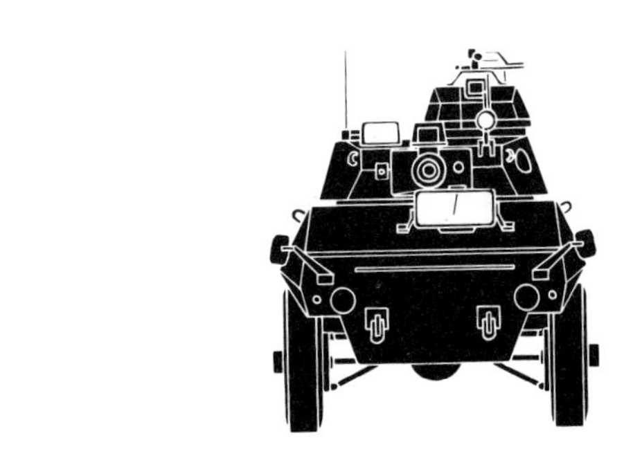

Role Armoured personnel carrier.

History Design work started in 1970, with a prototype being produced that year, and the first production vehicles were produced three years later. Uses similar components to Cascavel.

Employment Bolivia, Brazil, Chile, Colombia, Gabon, Guyana, Iraq, Libya, Tunisia, United Arab Emirates, Uruguay.

Vehicle data *Crew* 2 (commander, driver) + 13 infantrymen, *height* 2.45 m, *width* 2.44 m, *length* 5.76 m, *weight* 10,000 kg, *max. speed* 95 kph, *range* 600 km, *armament* variable (see below), *engine* Mercedes-Benz (Brazil) OM-352-A diesel developing 174 bhp at 2800 rpm.

Special characteristics Fully amphibious, using its wheels and hydrojets mounted at rear for propulsion with a max. speed of 10 kph. Night-driving aids. A winch can also be mounted. Optional NBC system.

Variants A wide number of weapon fits are available, including: pintle- and ring-mounted 7.62 mm or 12.7 mm mgs; 20 mm cannon (see silhouettes) and/or mgs mounted in turret; 81 mm and 60 mm breech-loading Brandt mortar (see photograph); 20 mm AA turret; Scorpion turret; 90 mm Cascavel turret (known as Hydracobra). Brazilian Marine version has four snorkel tubes let into the hull for use in heavy seas. Ambulance and command versions (with raised roof) and workshop variant also exist. Urutu 2 with Mercedes-Benz 190 hp, Urutu 3 with General Motors 212 hp and Urutu 4 also with 190 hp Mercedes-Benz are the basic models.

Manufacturer Engesa, São Paulo.

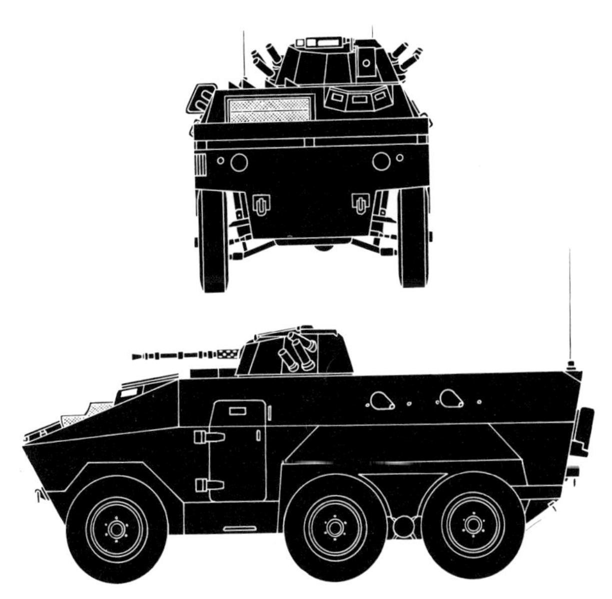

Role Armoured personnel carrier.

History This came into service with the Czech Army in 1964 as the Czech equivalent of the Russian BTR-50P, and is similar in appearance.

Employment Angola, Bulgaria, Czechoslovakia, Egypt, Hungary, India, Iraq, Israel, Libya, Morocco, Poland (known as TOPAS), Sudan.

Vehicle data *Crew* 2 (commander, driver) + 18 infantrymen, *height* 2.23 m, *width* 3.14 m, *length* 7.08 m, *weight* 15,000 kg, *max. speed* 58 kph, *range* 450 km, *armament* 1 × 82 mm recoilless gun (HEAT), 1 × 7.62 mm mg, *engine* PV-6 diesel developing 300 bhp at 1200 rpm.

Special characteristics Amphibious (using two water jets at max. speed of 10.8 kph), night-driving aids, NBC protection system.

Variants OT-62A (has no fixed armament); OT-62C (Polish variant mounting 14.5 mm and 7.62 mm mgs in a small turret; also has doors fitted into hull sides and has 3-man crew + 12 infantrymen); WPT-TOPAS (Polish recovery variant); command and ambulance versions.

Manufacturer Czech State Arsenals.

OT-62C

OT-64C (2)

Role Armoured personnel carrier.

History Designed jointly by the Czechs and Poles as an alternative to the Soviet BTR-60. Development began in 1959, and by 1964 it had entered service in large numbers. Chassis is based on the TATRA T813 8-ton truck.

Employment Czechoslovakia, Egypt, Hungary, India, Iraq, Libya, Morocco, Poland, Sudan, Syria, Uganda.

Vehicle data *Crew* 2 (commander, driver) + 18 infantrymen, *height* 2.03 m, *width* 2.5 m, *length* 7.44 m, *weight* 14,300 kg, *max. speed* 95 kph, *range* 710 km, *armament* 1 × 7.62 mm mg, *engine* Tatra T 928-14 diesel developing 180 bhp at 2000 rpm.

Special characteristics Amphibious, propelled by two propellers mounted at rear of hull with max. speed of 9 kph; NBC protection system; night-driving aids; front-mounted external winch.

Variants OT-64A (open turret); OT-64B (used only by Poland, and mounting 12.7 mm or 7.62 mm mg in open turret); OT-64C (1) (fully enclosed turret with 14.5 + 7.62 mm mgs); OT-64C (2) (OT-64C (1) with increased elevation for air defence); OT-64 ATGW vehicle using two Sagger launchers; OT-64 R-2 and R-3 ACV with heightened fighting compartment and 4 aerial bases; WPT SKOT ARV with crane. Czech M 1980 wheeled SP 152 mm howitzer is a derivative.

Manufacturer Czech State Arsenals.

OT-64A

AMX-13 LIGHT TANK

Role Light tank, also used for reconnaissance and as a tank destroyer.

History Design work started in 1946, with the first prototype being completed in 1948–9. It entered production in 1952, and came into service with the French Army the following year. Production is now complete.

Employment Algeria, Argentina, Chile, Dominican Republic, Ecuador, El Salvador, France, Indonesia, Ivory Coast, Jibuti, Lebanon, Morocco, Nepal, Netherlands, Peru, Singapore, Tunisia, Venezuela.

Vehicle data *Crew* 3 (commander, driver, gunner), *height* 2.3 m, *width* 2.5 m, *length* 4.88 m (6.36 m with gun forward), *weight* 15,000 kg, *max. speed* 60 kph, *range* 350–400 km, *armament* 1 × 90 mm (APDSFS, HEAT),

1 × 7.62 mm or 7.5 mm coaxial mg, *engine* SOFAM 8 GXb 8250 cc petrol developing 250 bhp at 3200 rpm.

Special characteristics Night-fighting and -driving equipment. Oscillating turret with autoloader.

Variants Model 51 (mounting the original 75 mm gun – see photograph); Dutch version with 105 mm gun in FL-2 turret; ATGW version mounting 2 × SS-11 and on later models HOT missile launchers on either side of turret; AMX-13 bridge-layer (folding Class 25 bridge); AMX-VC1 APC; AMX ARV. Both Austrian SK 105 and Brazilian Securi tank destroyers have AMX-13 turrets. Diesel versions also exist.

Manufacturer Creusot-Loire.

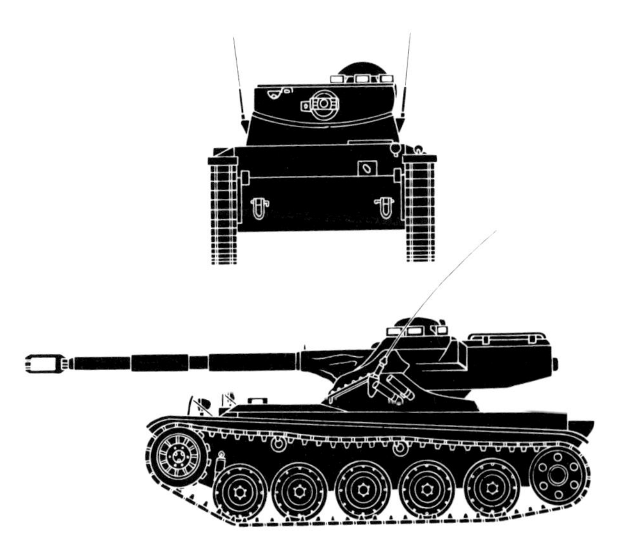

AMX-30 MAIN BATTLE TANK

Role Main battle tank.

History In the mid 1950s it was the intention to design a joint Franco-German MBT. AMX-30 was the French version, the first prototype being completed in 1960. The joint project did not materialize, and AMX-30 entered service with the French Army in 1966–7.

Employment Chile, France, Greece, Qatar, Saudi Arabia, Spain, United Arab Emirates, Venezuela.

Vehicle data *Crew* 4 (commander, loader/operator, gunner, driver), *height* 2.85 m, *width* 3.1 m, *length* 6.59 m, *weight* 36,000 kg, *max. speed* 65 kph, *range* 650 km, *armament* 1 × 105 mm (APDSFS, HEAT), 1 × 12.7 mm coaxial mg, 1 × 7.62 mm AA mg, *engine* Hispano-Suiza HS–110 multi-fuel developing 700 bhp at 2400 rpm.

Special characteristics NBC protection system; night-fighting and -driving aids. Can wade without preparation to a depth of 2 m, and 4 m using a snorkel. Optical rangefinder.

Variants AMX-30D ARV; AMX-30 bridgelayer; Pluton (tactical nuclear rocket) on AMX-30 chassis; AMX-30 401A AA tank with twin 30 mm cannon; AMX-30 with Roland AA missile system. AMX-30S hot climate export version; AMX-30B2 with COTAC integrated fire control system and laser rangefinder; AMX-30 Shahine with 6 Crotale AA missile launchers; Pluton SSM system. Diesel versions also exist.

Manufacturer ATS Roanne.

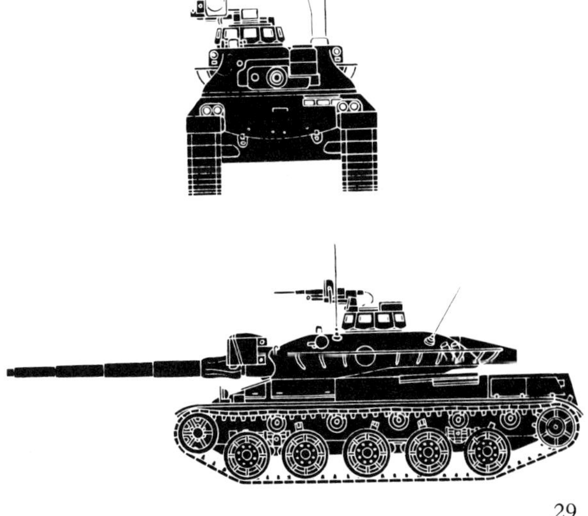

Role Armoured reconnaissance vehicle.

History Based on the 1938 AML201 prototype, development started shortly after the end of World War II, and the first prototype appeared in 1948. Production ran for ten years from 1950. It is now being replaced in the French Army by AMX-10RC.

Employment France, Mauritania, Morocco, Portugal, Tunisia.

Vehicle data *Crew* 4 (commander, gunner, 2 drivers), *height* 2.58 m, *width* 2.42 m, *length* 5.56 m (7.33 m with gun fully forward), *weight* 15,200 kg, *max. speed* 105 kph, *range* 650 km, *armament* 1 × 90 mm low pressure (APDSFS, HEAT), 2 × 7.62 mm mgs (coaxial and hull-mounted), *engine* Panhard 12-cylinder petrol developing 200 bhp at 3700 rpm.

Special characteristics Two driving positions, front and rear, and two pairs of retractable cross-country wheels. Night-viewing devices.

Variants Earlier models mounted 75 mm gun and 2 × 7.5 mm mgs. One version mounts AMX-13 light tank turret with 75 mm gun (see photograph). EBR ETT APC is in service in Portugal.

Manufacturer Panhard et Levassor.

Role Armoured car.

History Developed by Panhard in the late 1950s; the first prototype was produced in 1959, and the vehicle entered production in 1961.

Employment Algeria, Angola, Argentina, Bahrein, Burundi, Chad, Congo, Ecuador, El Salvador, France, Gabon, Iraq, Ireland, Ivory Coast, Jibuti, Kenya, Malaysia, Mauritania, Morocco, Niger, Nigeria, Portugal, Rwanda, Saudi Arabia, Senegal, Somalia, South Africa (known as Eland Mk V), Spain, Togo, Tunisia, United Arab Emirates, Upper Volta, Venezuela, Yemen, Zaire, Zimbabwe.

Vehicle data *Crew* 3 (commander, gunner, driver), *height* 2.07 m, *width* 1.97 m, *length* 3.79 m, *weight* 5500 kg, *max. speed* 100 kph, *range* 600 km, *armament* 1 × 90 mm gun, 1 × 7.62 mm coaxial mgs, *engine* Panhard 4HD petrol developing 90 bhp at 4700 rpm.

Special characteristics Night-fighting and -driving aids; amphibious using flotation screen (max. speed 7 kph).

Variants AML with HE60–7 turret (60 mm mortar, 2 × 7.62 mm mgs) – see silhouettes; AML with HE60–12 turret (12.7 mm mg and 60 mm mortar), with HE60–20 turret (20 mm cannon and 60 mm mortar); AML 30 (with 30 mm cannon); with S530 turret (2 × 20 mm AA cannon); ATGW version with 4 × ENTAC ATGW launchers, EPR (turretless with 12.7 mm mg); AML20 (20 mm cannon, 7.62 mm mg).

Manufacturer Panhard et Levassor; Sandoek-Austral Ltd, South Africa.

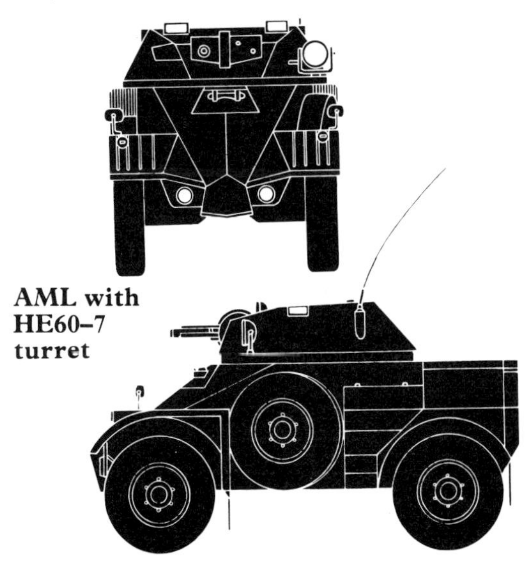

AML with HE60–7 turret

Role Armoured reconnaissance vehicle.

History This vehicle was developed by Renault especially for the export market, and made its first appearance in 1979. It is, however, used by the Garde Républicaine. Chassis is the same as used on the Saviem-Creusot-Loire VAB APC. (Saviem is now part of Renault.)

Employment France, Oman, United Arab Emirates.

Vehicle data *Crew* 3 (commander, gunner, driver), *height* 2.55 m, *width* 2.49 m, *length* 5.50 m (8.15 m with gun fully forward), *weight* 12,800 kg, *max. speed* 92 kph, *range* in excess of 1000 km, *armament* 1 × 90 mm HV rifled gun (APDSFS, HEAT, HE, Canister, Smoke),

1 × 7.62 mm coaxial mg, *engine* MAN D.2356 air-cooled diesel developing 235 bhp at 2200 rpm.

Special characteristics Night vision and fighting equipment, NBC protection, self-towing winch, amphibious (using flotation screen).

Variants Among other weapon fits on offer is the Belgian Cockerill 90 mm low-pressure gun, 12.7 mm Browning as a coaxial mg, and a turret roof-mounted 7.62 mm (or Browning 12.7 mm) mg, 81 mm smoothbore gun/mortar, 2 × 20 mm AA cannon. Laser rangefinder and power-assisted traverse are also available.

Manufacturer Société des Matériels Spéciaux Renault VI – Creusot-Loire.

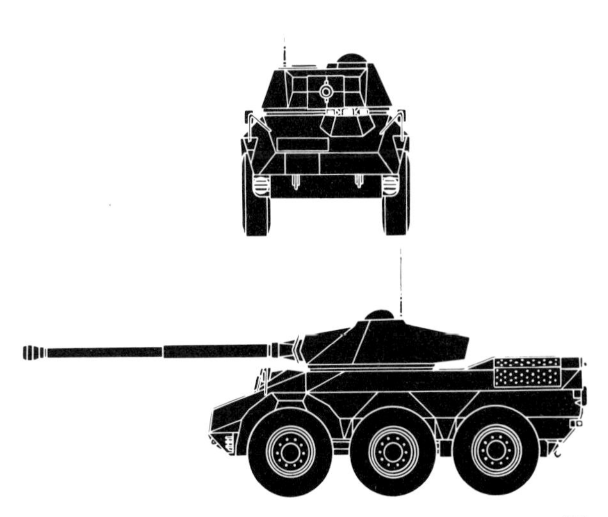

PANHARD ERC 90 SAGAIE

Role Armoured reconnaissance vehicle.

History As with the Renault VBC 90, this vehicle has been developed specifically for the export market, although it is employed with the French Force d'Intervention. Prototypes first appeared in 1977 and production began in 1979. Turret is the same as Renault's VBC 90.

Employment Argentina (known as Lynx), Chad, France, Iraq, Ivory Coast, Mexico (Lynx), Niger.

Vehicle data *Crew* 3 (commander, gunner, driver), *height* 2.24 m, *width* 2.50 m, *length* 5.02 m (7.79 m with gun fully forward), *weight* 7800 kg, *max. speed* 110 kph, *range* 950 km, *armament* 90 mm low-pressure gun (HEAT, HE, Smoke), 1 × 7.62 mm coaxial and 1 × 7.62 mm turret-mounted AA mg, *engine* Peugeot V-6 petrol developing 140 bhp at 5250 rpm.

Special characteristics Centre pair of road wheels is retractable and not normally used on roads. Amphibious capability is available – max. speed 4.5 kph propelled by its own wheels and 9.5 kph using water jets. Night-viewing and -fighting aids, as well as NBC protection.

Variants ERC60-20 (with 60 mm breech-loading mortar and 20 mm cannon), ERC 90 Lynx as employed in Argentina and Mexico with 90 mm gun in Lynx turret, 60 mm mortar versions, 2 × 20 mm cannon, Hughes 25 mm Chain Gun (Lanza), SATCP SAM missile launcher.

Manufacturer Panhard et Lavassor.

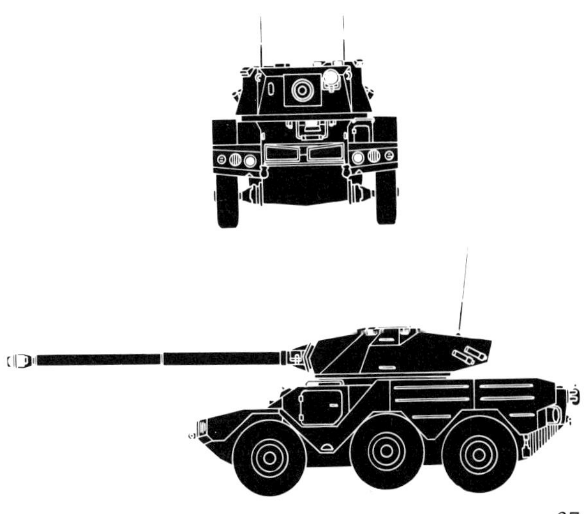

Role Armoured reconnaissance vehicle.

History Designed to replace the ageing Panhard EBR armoured car, AMX-10RC began to enter service with the French Army in 1981 – the first prototype having been produced in 1973. It shares many common components with AMX-10P APC.

Employment France, Morocco.

Vehicle data *Crew* 4 (commander, gunner, radio operator, driver), *height* 2.565 m, *width* 2.84 m, *length* 6.243 m (hull only), *weight* 15,000 kg, *max. speed* 85 kph, *range* 800 km, *armament* 1 × 105 mm low-pressure (APDSFS, HEAT), 1 × 7.62 mm coaxial mg, *engine* Hispano-Suiza HS-115 diesel developing 280 bhp at 3000 rpm.

Special characteristics Fully amphibious using hydro-jet propulsion with a max. speed of 7.2 kph; NBC protection system; night-fighting and -driving aids.

Variants AMX–10C, which is a tracked version using same hull and turret, as yet only in prototype. AMX-10RAA version with 2 × 20 mm cannon, AMX-10 RTT APC (Crotale 12.7 mm mg turret to rear, crew 3 + 10).

Manufacturer ATS Roanne.

Role Light airportable ATGW and scout vehicle.

History In 1978 the French Army drew up a requirement for a very light airportable ATGW and scout vehicle for use with airmobile and airborne forces. In 1985 Panhard was awarded a production contract and the first vehicles entered service at the end of 1986.

Employment France, Mexico.

Vehicle data *Crew* 3 (commander, driver, missile controller), *height* 1.7 m (to hull top), *width* 2.02 m, *length* 3.7 m, *weight* 3,590 kg, *max. speed* 100 kph, *range* 750 km, *armament* Milan ATGW launcher, 1 × 7.62 mm mg, *engine* Peugeot XD 3T developing 105 bhp at 4150 rpm.

Special characteristics NBC protection system and amphibious with max. water speed of 4 kph.

Variants The scout version has a 2-man crew and is armed with 1 × 7.62 mm or 1 × 12.7 mm mg. Numerous other variants are proposed.

Manufacturer Panhard et Levassor SA.

155 MM SP HOWITZER MK F3

Role Self-propelled artillery howitzer.

History Developed in the early 1960s, it entered service in the mid 1960s. It is being replaced in the French Army by AMX 155 mm SP GCT. It uses the AMX-13 chassis.

Employment Argentina, Chile, Ecuador, France, Kuwait, Morocco, Qatar, Sudan, United Arab Emirates, Venezuela.

Vehicle data *Crew* 2 (commander and driver), remaining crew members are carried separately, *height* 2.1 m, *width* 2.72 m, *length* 6.22 m, *weight* 17,400 kg, *max. speed* 65 kph, *range* 300 km, *armament* 1 × 155 mm, *engine* SOFAM 8 GXb petrol developing 250 bhp at 3200 rpm.

Special characteristics Night-driving aids. Is accompanied by AMX-13 VCI, which carries remainder of crew and ammunition. Can use US M107 155 mm SP ammunition.

Variants Diesel engine version.

Manufacturer Creusot-Loire, Chalon-sur-Saône.

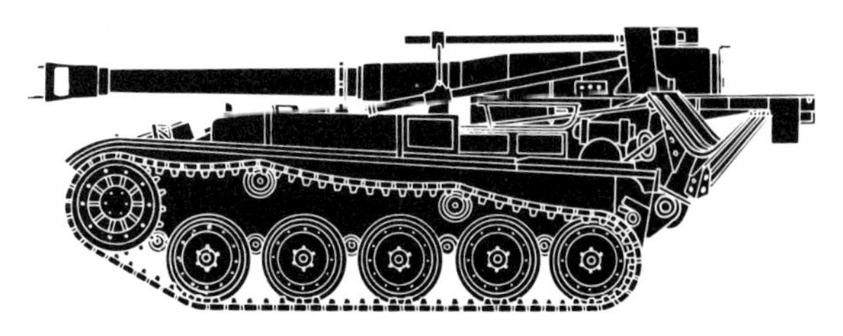

Role Self-propelled artillery gun.

History Design work started in 1969 on this replacement for both the SP 155 mm howitzer Mk F3 and the AMX 105 mm SP gun. Prototypes first appeared in public in 1973, and it entered production in 1977, coming into service with the French Army shortly afterwards. It uses the AMX-30 chassis.

Employment France, Iraq, Saudi Arabia.

Vehicle data *Crew* 4 (commander, driver and 2 gunners), *height* 3.3 m, *width* 3.15 m, *length* 10.4 m (with gun fully forward), *weight* 41,000 kg, *max. speed* 60 kph, *range* 450 km, *armament* 1 × 155 mm, 2 × 7.62 mm AA mgs, *engine* Hispano-Suiza HS-110 multi-fuel developing 700 bhp at 2400 rpm.

Special characteristics NBC protection system and night-driving aids. Autoloader giving rate of fire of 8 rds/min.

Variants Nil.

Manufacturer ATS Roanne.

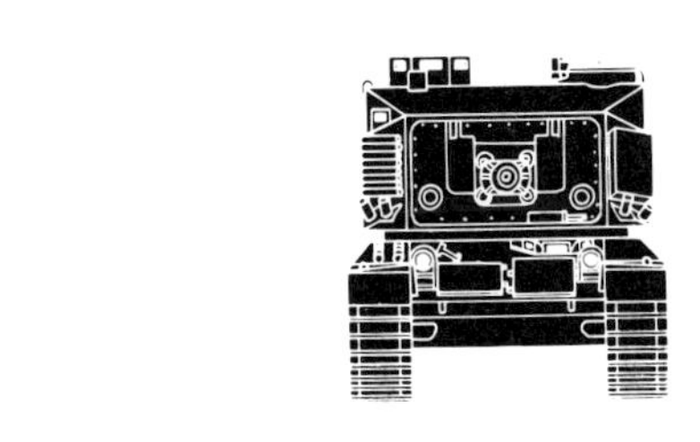

Role Mechanized infantry combat vehicle.

History Design commenced in the early 1960s as successor to AMX-13 VCI and prototypes were first produced in 1968. It entered service with the French Army in 1973.

Employment France, Greece, Indonesia, Mexico, Morocco, Qatar, Saudi Arabia, United Arab Emirates.

Vehicle data *Crew* 2 (commander, driver) + 9 infantrymen, *height* 2.54 m, *width* 2.78 m, *length* 5.78 m, *weight* 13,800 kg, *max. speed* 65 kph, *range* 600 km, *armament* 1 × 20 mm cannon, 1 × 7.62 mm coaxial mg, *engine* Hispano-Suiza HS-115-2 diesel developing 276 bhp at 3000 rpm.

Special characteristics Amphibious (propelled by

water jets with max. speed of 8 kph); NBC protection system; night-fighting and driving aids.

Variants AMX-10P 25 MICV (25 mm cannon); AMX-10 SAO and VOA Artillery Observation Vehicles; Artillery fire control vehicle (using RATAC radar); HOT ATGW vehicle (with 4 × HOT launchers); AMX-10PC (ACV); AMX-10 ECH (ARV); ambulance; AMX-10TM, towing 120 mm Brandt mortar; AMX-10 TM81 with 81 mm mortar; AMX-10 PAC 90 with 90 mm gun (Indonesia only); Artillery OP version with 2-man turret. Yugoslav M-980 MICV uses many AMX-10P components.

Manufacturer ATS Roanne.

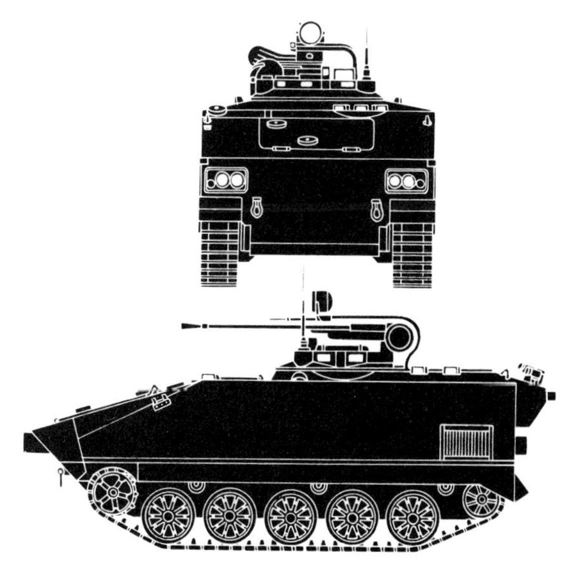

SAVIEM/ CREUSOT-LOIRE VAB

Role Armoured personnel carrier.

History Developed to supplement AMX-10P MICV: design work started in 1969, with the first prototypes appearing in 1973. Production commenced in 1977. Uses many commercial components. VBC 90 uses same chassis.

Employment Central African Republic, Cyprus, France, Ivory Coast, Lebanon, Mauritius, Morocco, Oman, Qatar, United Arab Emirates.

Vehicle data *Crew* 2 (commander, driver) + 10 infantrymen, *height* 2.06 m, *width* 2.49 m, *length* 5.98 m, *weight* 13,000 kg, *max. speed* 100 kph, *range* 1000 km, *armament* pintle-mounted 7.62 mm or 12.7 mm mg, or turret-mounted 20 mm cannon, *engine* Saviem HM-71 2356 diesel developing 230 bhp at 2220 rpm.

Special characteristics Amphibious (propelled by 2 water jets at max. speed of 7 kph); NBC protection system; night-driving aids.

Variants 4 × 4 version exists (see silhouettes) and is known as VIB by French Air Force. Other variants are ambulance; two anti-tank versions mounting 4 × HOT ATGW launchers; ARV; ACV; Internal Security version with dozer blade; surveillance with RATAC radar; 120 mm mortar tractor; 81 mm mortar version; VCI (IFV with 20 mm cannon turret); repair; VCAC Milan; ATGW; VAC PC command; engineer vehicle; 2 × 20 mm AA cannon.

Manufacturer Saviem/Creusot-Loire.

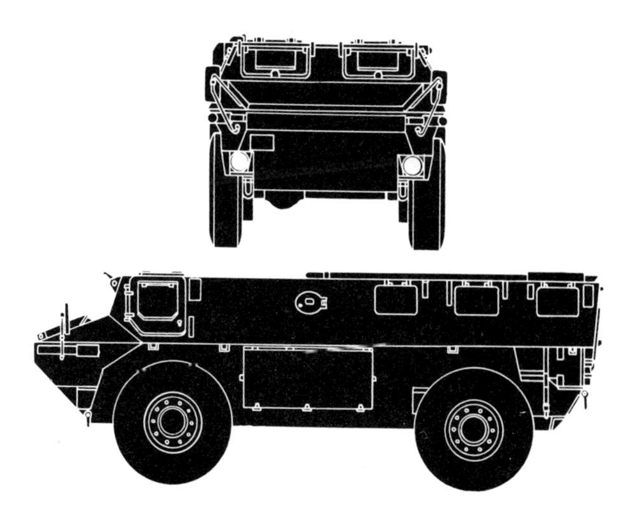

Role Self-propelled anti-aircraft gun system.

History This was a project involving a number of French and Swiss firms, begun in 1960. The first prototypes were completed in 1962, and the system entered service with the French Army in 1965. It uses the AMX-13 chassis, with French radar and Swiss-designed armament.

Employment France, Saudi Arabia.

Vehicle data *Crew* 3 (commander, gunner, driver),

height 3.794 m (with radar operational), *width* 2.5 m, *length* 5.373 m, *weight* 17,200 kg, *max. speed* 60 kph, *range* 300 km, *armament* 2 × 30 mm cannon, *engine* SOFAM 8 GXb petrol developing 250 bhp at 3200 rpm.

Special characteristics Uses RD515 Oeil Noir target-acquisition/tracking radar with folding antenna, which also acts as rangefinder.

Variants Saudi-Arabia version AMX-30 SA has new turret with improved radar, AMX-30 S (2 × 30 mm Sabre turret).

Manufacturer Creusot-Loire.

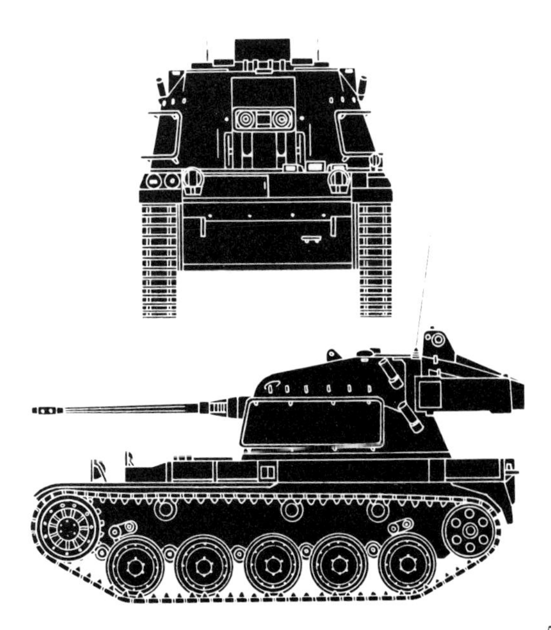

Role Main battle tank.

History Design commenced in 1955 as a parallel exercise to the French AMX-30. First prototype was produced in 1960, and production proper started in 1965, entering service with FRG the following year. Many NATO armies purchased Leopard rather than its British contemporary, Chieftain, because it was cheaper and simpler for conscripts to operate.

Employment Australia, Belgium, Canada, Denmark, Federal Republic of Germany, Greece, Italy, Netherlands, Norway, Turkey.

Vehicle data *Crew* 4 (commander, gunner, loader/operator, driver), *height* 2.64 m, *width* 3.25 m, *length* 7.09 m (9.54 m with gun over front), *weight* 40,000 kg, *max. speed* 65 kph, *range* 600 km, *armament* 1 × 105 mm rifled (APDS, APDSFS, HEAT, HESH), 1 × 7.62 mm coaxial mg, 1 × 7.62 mm AA mg, *engine* MTU MB 838 Ca. M500 multi-fuel developing 830 bhp at 2200 rpm.

Special characteristics Night-fighting and -driving aids; NBC protection system. Wades to 2.25 m without preparation, and 4 m with snorkel. 105 mm is British-designed standard NATO gun. Stereoscopic rangefinder but Australian, Belgian, Canadian versions have laser rangefinder.

Variants Leopard 1A1 with spaced armour turret; 1A1A1 (appliqué armour on turret); 1A2 (no special armour on turret); 1A3 (modified turret, with wedge-shaped mantlet); 1A4 (with IFCS); ARV; AEV (with dozer blade and auger); bridgelayer. Gepard AA system uses same chassis (see separate entry).

Manufacturer Krauss-Maffei; OTO Melara (Italy).

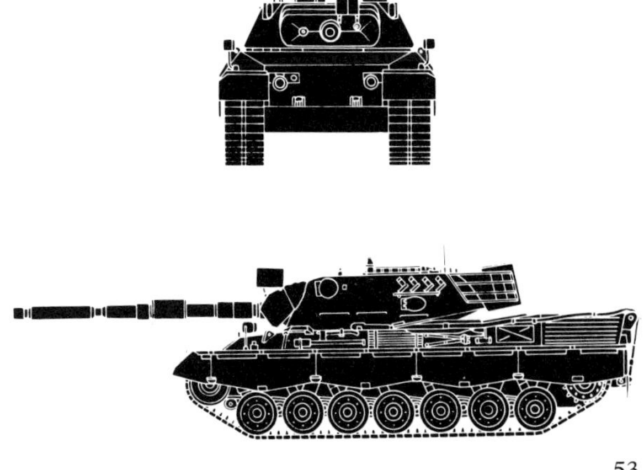

LEOPARD 2 (KAMPFPANZER 2)

Role Main battle tank.

History Development started in the late 1960s, but on low priority until the cancellation of the joint US/FRG main battle tank project in 1970. 17 prototypes were produced, of which one went to the USA for comparison trials with AMX-1. Production commenced in 1979, and it entered service with the Bundeswehr the following year.

Employment Federal Republic of Germany, Netherlands, Switzerland.

Vehicle data *Crew* 4 (commander, gunner, loader/operator, driver), *height* 2.49 m, *width* 3.54 m, *length* 7.73 m (9.74 m with gun forward), *weight* 50,000 kg, *max. speed* 68 kph, *range* 800 km, *armament* 1 × 120 mm smooth-bore (APDSFS, HEAT) 1 × 7.62 mm coaxial, 1 × 7.62 mm AA mg, grenade launcher, *engine* turbocharged MTU MB 873 Ka-500 multi-fuel developing 1500 bhp at 2600 rpm.

Special characteristics Night-fighting and -driving aids; NBC protection system. Integrated fire-control system with laser rangefinder.

Variants Nil at present, but ARV and AEV under development.

Manufacturer Krauss-Maffei.

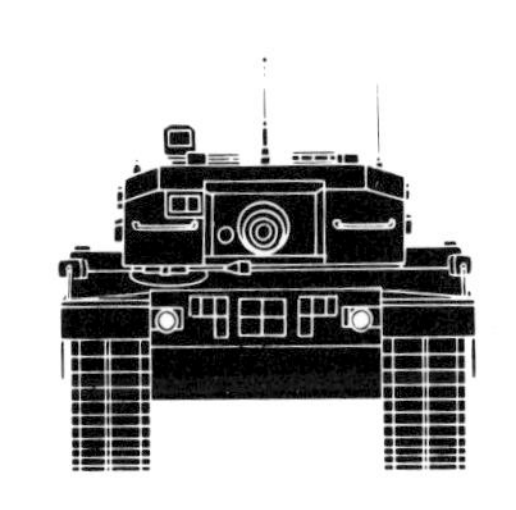

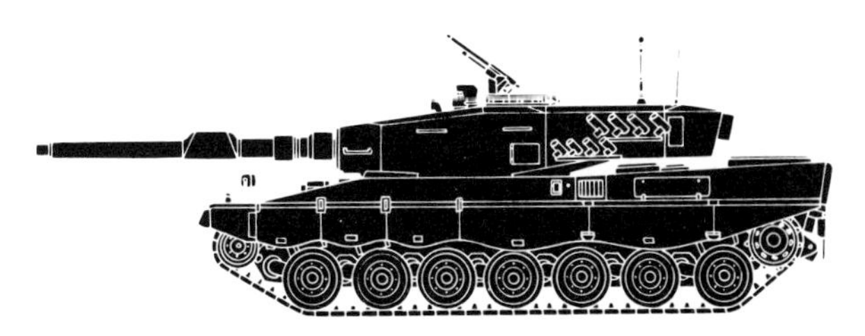

Role Medium battle tank.

History In 1974 Thyssen Henschel were awarded a contract to design and develop a medium tank for Argentina to be known as TAM. The first prototype appeared in 1976 and deliveries to Argentina began in 1980. In the meantime, Thyssen Henschel carried out futher development of the TAM, with a more powerful engine and improved fire control system, and this is known as TH 301. The TAM/TH 301 chassis is based on the Marder MICV (see separate entry).

Employment Argentina, Panama, Peru (all TAM).

Vehicle data *Crew* 4 (commander, gunner, driver, loader), *height* 2.436 m, *width* 3.306 m, *length* 6.62 m (8.17 m gun fully forward), *weight* 30,500 kg, *max. speed* 72 kph, *range* 750 km, *armament* 1 × 105 mm rifled gun (HEAT, HESH, APDS, APDSFS), 2 × 7.62 mm mgs (coaxial and AA), *engine* MTU diesel MB 833 KA 500 developing 750 bhp at 2400 rpm.

Special characteristics Integrated fire control system incorporating laser rangefinder and night vision equipment including low light level TV.

Variants VICP MICV with 2–man turret (20 mm cannon and 7.62 mm mg) uses the same TAM/Marder chassis and is also used by Argentina.

Manufacturer Thyssen Henschel.

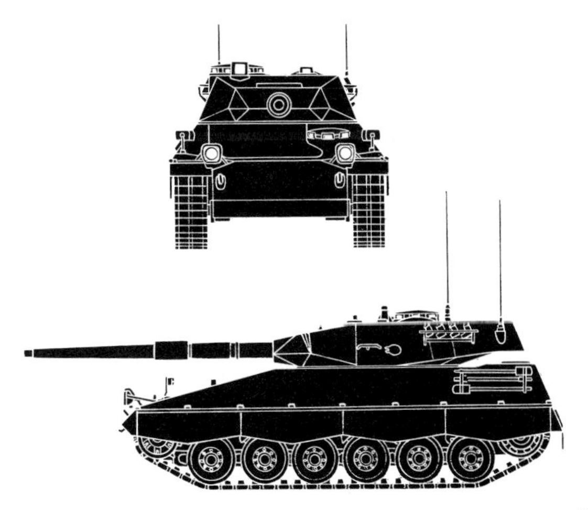

Role Reconnaissance vehicle.

History Development began in 1964. Intended as a replacement for SPz 11-2, the first prototypes appeared in 1967–8, and Luchs entered production in 1975. Production was completed in 1978.

Employment Federal Republic of Germany.

Vehicle data *Crew* 4 (commander, gunner, 2 drivers), *height* 2.84 m, *width* 2.98 m, *length* 7.743 m, *weight* 19,500 kg, *max. speed* 90 kph, *range* 800 km, *armament* 1 × 20 mm cannon, 1 × 7.62 mm commander's mg, *engine* Daimler-Benz OM 403VA developing 390 bhp (diesel) and 300 bhp (petrol) at 1800 rpm.

Special characteristics Fully amphibious, being driven by two propellers in rear of hull with max. speed of 10 kph; rear-facing driving position; NBC protection system; night-fighting and -driving aids.

Variants Nil, although there are proposals for fitting of Oerlikon 35 mm turret, and Roland SAM system.

Manufacturer Rheinstahl Wehrtechnik.

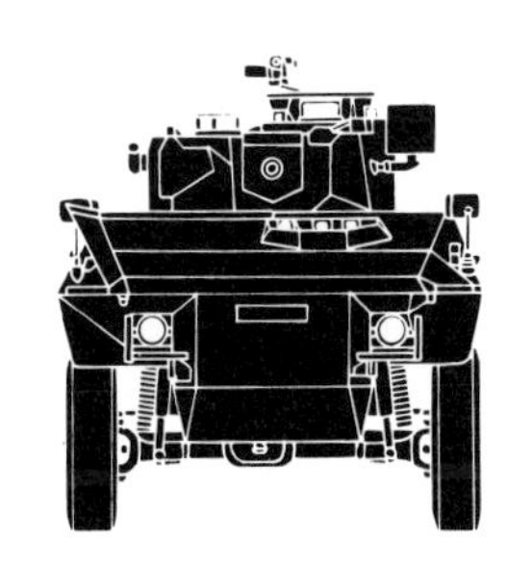

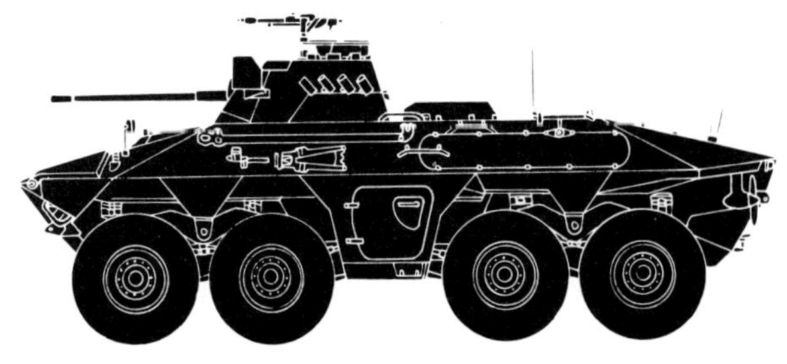

Role Mechanized infantry combat vehicle.

History In 1959 the Bundeswehr drew up a requirement for an MICV, and as a result three firms designed and built a number of prototypes over the period 1960–3. In 1969 Marder was cleared for production, and entered service with the Bundeswehr three years later.

Employment Argentina, Brazil, Federal Republic of Germany.

Vehicle data *Crew* 4 (commander, driver, 2 gunners) + 6 infantrymen, *height* 2.95 m, *width* 3.24 m, *length* 6.79 m, *weight* 28,200 kg, *max. speed* 75 kph, *range* 520 km, *armament* 1 × 20 mm cannon, 2 × 7.62 mm mg (coaxial and rear hull-mounted), *engine* MTU MB 833 Ea-500 diesel developing 600 bhp at 2200 rpm.

Special characteristics Night-driving and -fighting

aids; NBC protection system; can ford to depth of 2.5 m using snorkel.

Variants Milan ATGW launcher mounted on side of turret; Marder with 25 mm cannon; Radarpanzer TÜR (no turret but Siemens radar mounted on hydraulic arm); Waffenträger ROLAND (with two ROLAND SAM launchers), and is used by Brazil. Argentine VCTP has different turret. Argentine 105 mm TAM tank uses Marder chassis.

Manufacturer Rheinstahl Sonderfertigung, Kassel; Atlas Mak, Kiel.

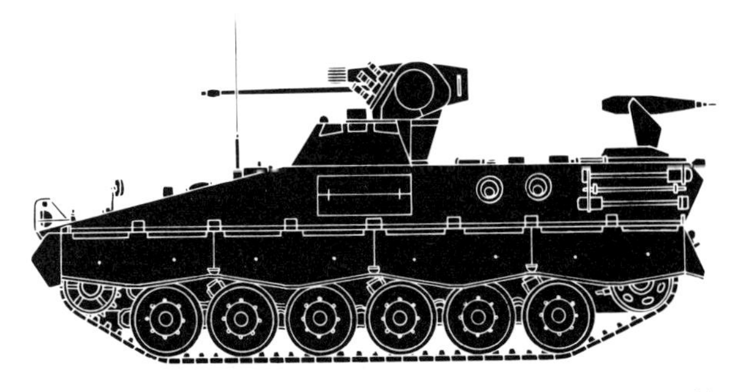

FRG **FUCHS (TRANSPORTPANZER 1)**

Role Multi-purpose armoured carrier.

History Fuchs was developed as a result of a Bundeswehr requirement drawn up in 1970 for a multi-purpose vehicle for HQs and other units which could survive artillery fire and air attack. Prototypes appeared in 1974 and it entered Army service in 1980. It has some common components with Luchs.

Employment Federal Republic of Germany, Venezuela.

Vehicle data (basic vehicle) *Crew* 2 (commander, driver) + 10 infantrymen, *height* 2.43 m, *width* 2.98 m, *length* 6.83 m, *weight* 16,200 kg, *max. speed* 105 kph, *range* 800 km, *armament* 1 × 7.62 mm mg, *engine* Mercedes-Benz Diesel OM 402A developing 320 bhp at 2500 rpm.

Special characteristics Fully amphibious using 2 rudder screws which give a speed of 10 kph, NBC protection system.

Variants A number of conversion kits exist: ambulance (San); AEV (Pi); NBC reconnaissance (ABC); command and communications (Fü Fu); surveillance with RASIT radar; electronic warfare (TPZl-Eloka). 8×8 version exists in prototype and both this and 6×6 can mount a turret with a variety of weapons fits ranging from mgs to 90 mm gun.

Manufacturer Thyssen-Henschel.

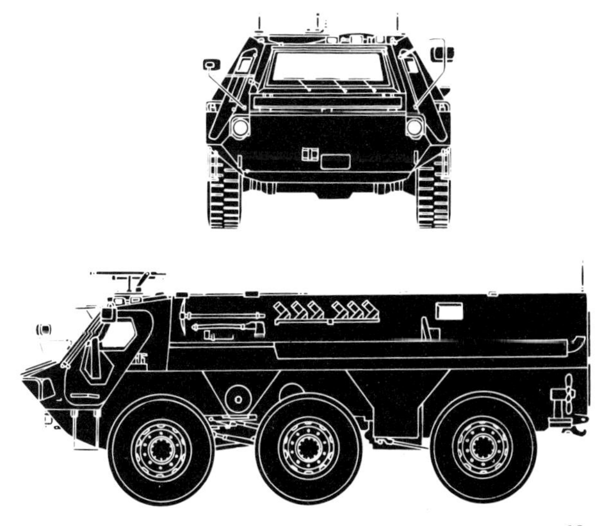

GEPARD (FLAKPANZER 1)

Role Self-propelled anti-aircraft gun.

History This resulted from a requirement put out by the Bundeswehr in the late 1960s. German and Swiss firms competed for the development contract, with the Swiss firm of Contraves winning it. It entered service with the Bundeswehr in 1978. It uses the Leopard 1 hull. Production was completed in 1980.

Employment Belgium, Federal Republic of Germany, Netherlands (known as Cheetah).

Vehicle data *Crew* 3 (commander, gunner, driver), *height* 4.03 m (with radar in operation), *width* 3.25 m, *length* 7.7 m (guns forward), *weight* 45,100 kg, *max. speed* 65 kph, *range* 600 km, armament 2 × 35 mm cannon, *engine* MTU MB 838 Ca. 500 multi-fuel developing 830 bhp at 2200 rpm.

Special characteristics NBC protection system. Has Siemens search radar and Siemens-Albis tracking radar.

Variants Cheetah uses a Dutch radar with a long narrow dish (This is integrated search/tracking Hollandse radar). TÜR armoured radar vehicle is a modified Gepard.

Manufacturer Krauss-Maffei.

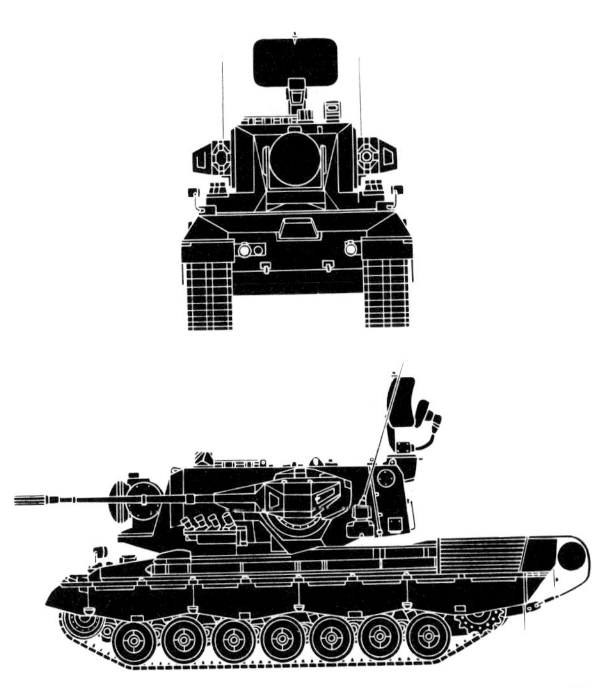

Role Airportable light armoured fighting vehicle.

History To meet a Bundeswehr requirement of 1971 for a tracked or wheeled AFV for use with airborne forces and small enough to be carried in the CH-53 helicopter, Porsche eventually won a development contract with their tracked concept. In 1978, however, after prototypes had been fielded, the contract was cancelled because of financial stringency. In 1984, however, after Porsche, now joined by Krupp, had continued development work privately, the Bundeswehr finally placed a production order. Weisel will be in Bundeswehr service by 1989.

Employment Not yet in service and then initially with Federal Republic of Germany.

Vehicle data *Crew* 3 (commander, gunner, missile operator), *height* 1.875 m, *width* 1.82 m, *length* 3.265 m, *weight* 2750 kg, *max. speed* 80 kph, *range* 200 km, *armament* 1 × TOW/HOT launcher, *engine* VW diesel developing 86 bhp at 2,000 rpm.

Special characteristics NBC protective system, amphibious with max. speed of 12 kph (propellers), night fighting and driving aids.

Variants The main variant is a 20 mm cannon version with just a 2-man crew, but there is also a TOW-2 variant (see silhouettes). Other proposals are ACV, radar, Stinger SAM, ARV versions.

Manufacturer Porsche and Krupp.

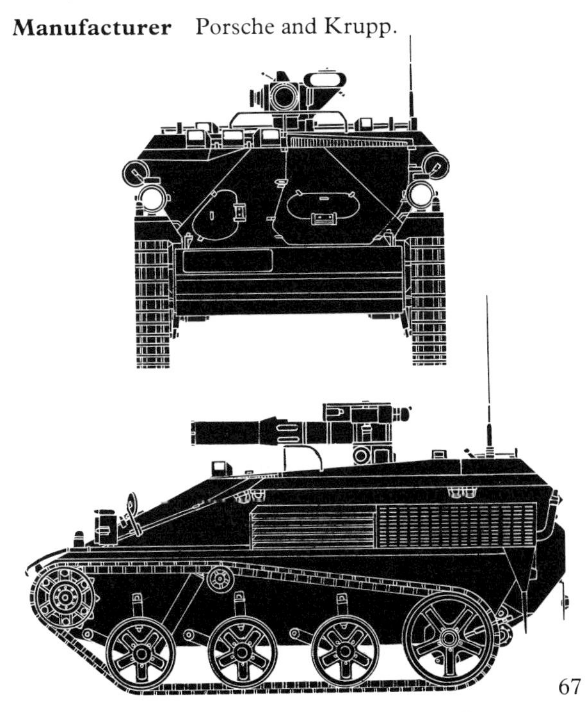

Role Amphibious reconnaissance vehicle.

History It was first noted in the late 1960s, and equates to the Russian BRDM. Entered service with Hungarian Army in 1964, and Czech and Polish armies in 1966.

Employment Czechoslovakia (known as OT-65), Hungary, Poland.

Vehicle data *Crew* 3 (commander, gunner, driver), *height* 2.525 m, *width* 2.362 m, *length* 5.79 m, *weight* 7000 kg, *max. speed* 100 kph, *range* 500 km, *armament* 1 × 14.5 mm, 1 × 7.62 mm coaxial mg, *engine* Csepel 4-cylinder diesel developing 100 bhp at 2300 rpm.

Special characteristics Amphibious, with propulsion from 2 water jets at rear of hull with max. speed of 10 kph; night-fighting and -driving aids; NBC protection system. Has 2 × 2 retractable cross-country wheels.

Variants Turretless version; OT-65A with turret of

OT-62B APC with 7.62 mm mg mounted and 82 mm externally mounted recoilless gun; NBC reconnaissance version. A later derivative is the PSzH-IV APC, which has 14.5 mm turret-mounted mg, no belly wheels and doors in hull sides (see silhouettes). Ambulance and command versions exist of this.

Manufacturer Hungarian State Arsenals.

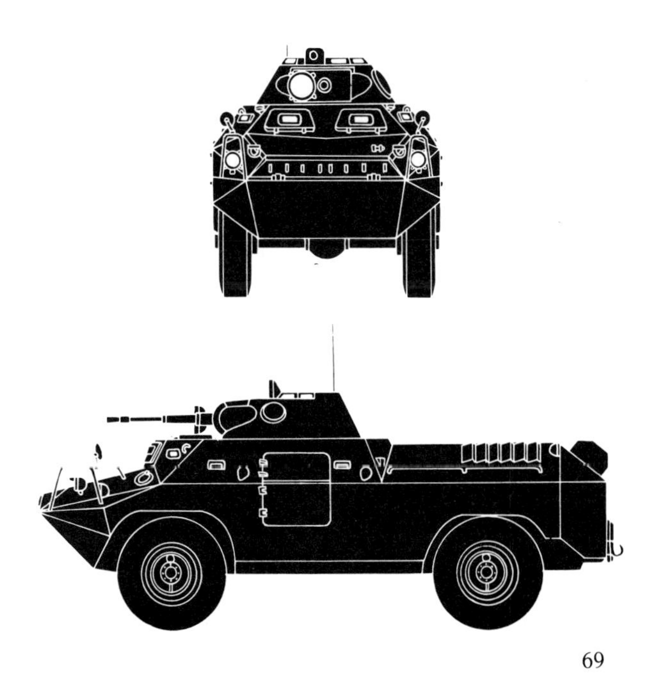

Role Main battle tank.

History As a result of the 1967 Arab–Israeli War experience, the decision was made in 1970 to develop an MBT with the emphasis on protection. First prototype appeared in 1972, and it entered service in 1979.

Employment Israel.

Vehicle data *Crew* 4 (commander, gunner, loader/radio operator, driver), *height* 2.64 m, *width* 3.70 m, *length* 7.45 m (8.63 m with gun fully forward), *weight* 56,000 kg, *max. speed* 44 kph, *range* 400 km, *armament* 1 × 105 mm rifled gun (APDSFS, APDS, HEAT, HESH, Canister), 3 × 7.62 mm mgs (one coaxial, two AA on turret roof), 1 × 60 mm mortar mounted on turret roof, *engine* Teledyne Continental Motors AVDS-1790-5A diesel developing 908 bhp at 2400 rpm.

Special characteristics Unlike most MBTs, engine is at the front. Rear hull has a compartment which can be used for ammunition stowage, to carry infantrymen, as a command post or ambulance. Rear doors also enable quick crew evacuation. Night-vision and -fighting equipment, NBC protection system. Fire-control system incorporates laser rangefinder.

Variants Mk 2 with special armour and mortar mounted inside turret. Mk 3 with hydropneumatic suspension, new power pack, 120 mm smoothbore gun. Most now have Blazer appliqué armour around turret.

Manufacturer Israeli Ordnance Corps.

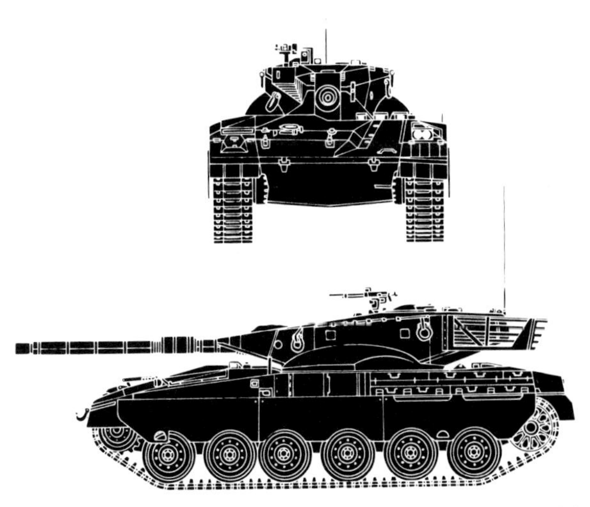

Role Reconnaissance vehicle.

History Entered service with the Israeli Defence Forces in 1975.

Employment Guatemala, Honduras, Israel.

Vehicle data *Crew* 8 (commander, driver, 6 riflemen), *height* 1.66 m, *width* 2.03 m, *length* 5.023 m, *weight* 4000 kg, *max. speed* 100 kph, *range* 550 km, *armament* 4 × 7.62 mm or 4 × 12.7 mm mgs, *engine* Dodge 225.2 petrol developing 120 bhp at 3000 rpm.

Special characteristics A winch can be mounted.

Variants 106 mm recoilless gun or 2 × 20 mm AA cannon can be mounted. Development of RBY is the RAM V-1 with same chassis, but diesel engine and higher hull sides and overhead protection. A number of variants of this have been developed.

Manufacturer Ramta Structures and Systems.

Role Self-propelled 155 mm howitzer.

History This adaptation of the Sherman tank chassis was developed in the early 1970s, and first made its appearance in significant numbers during the 1973 Arab–Israeli War.

Employment Israel.

Vehicle data *Crew* 8, *height* 2.46 m, *width* 3.33 m, *length* 8.55 m (including gun), *weight* 41,500 kg, *max. speed* 37 kph, *range* 260 km, *armament* 1 × 155 mm, 1 × 7.62 mm mg, *engine* Cummins VT 8-460-Bi diesel developing 300 bhp at 2800 rpm.

Special characteristics Nil.

Variants Nil. Soltam have designed a new 155 mm gun fitted on a Centurion chassis, and also suitable for other MBT chassis.

Manufacturer Soltam Company, Israel.

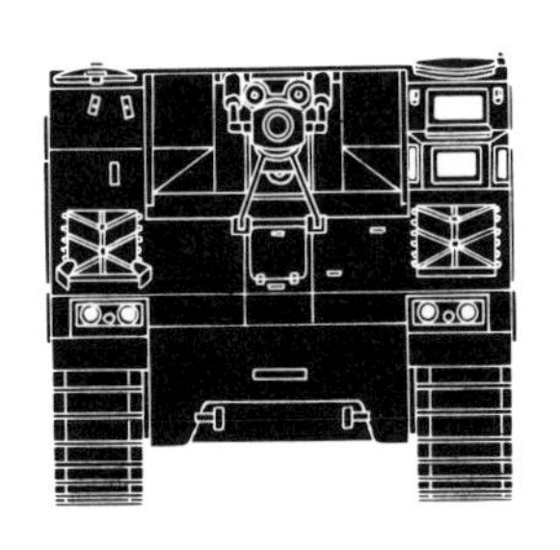

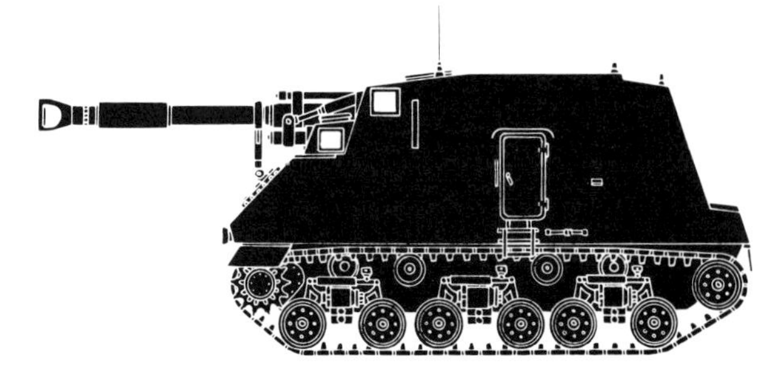

Role Main battle tank.

History Development by OTO–Melara began in 1977 as a private venture for the export market. A prototype appeared in 1980 and production began the following year. Hull and some automotive components very similar to Leopard 1.

Employment United Arab Emirates.

Vehicle data *Crew* 4 (commander, gunner, loader, driver), *height* 2.42 m, *width* 3.51 m, *length* 8.33 m (9.65 m with gun fully forward), *weight* 43,000 kg, *max. speed* 65 kph, *range* 600 km, *armament* 1 × 105 mm rifled gun (APDS, HEAT, HESH), 2 × 7.62 mm mg (coaxial and AA), *engine* Fiat V-10 supercharged multi-fuel developing 830 bhp at 2200 rpm.

Special characteristics Can wade up to depths of 2.25 m with little preparation, and up to 4 m with additional snorkel equipment. Night-fighting and -viewing equipment – LLTV is also available. NBC protection system, and integrated fire-control system including laser rangefinder.

Variants Mk 2 with Galileo IFCS; ARV (for UAE). Chassis is being used for the OTO–Melara Palmaria 155 mm SP howitzer, which is now in production for export market (see separate entry).

Manufacturer OTO–Melara.

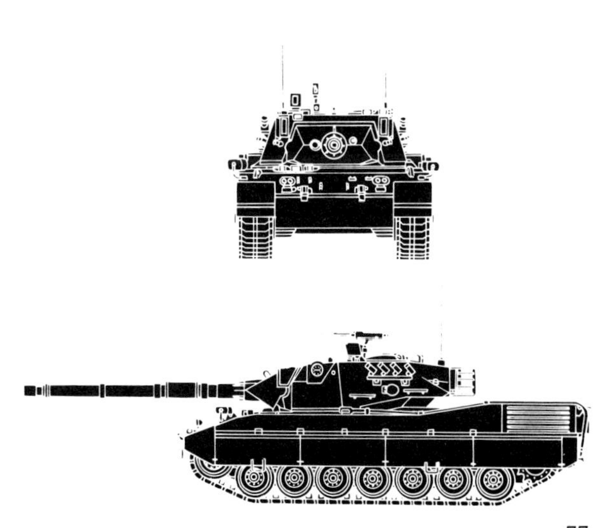

FIAT-MELARA TYPE 6616

Role Armoured car.

History A joint development between Fiat and OTO–Melara, the first prototype was produced in 1972 and, after trials, it entered service with the Italian carabinieri.

Employment Italy, Peru, Somalia

Vehicle data *Crew* 3 (commander, gunner, driver), *height* 2.035 m, *width* 2.5 m, *length* 5.37 m, *weight* 8000 kg, *max. speed* 100 kph, *range* 700 km, *armament* 1 × 20 mm cannon, 1 × 7.62 mm coaxial mg, *engine* Model 8062.24 diesel developing 160 bhp at 3200 rpm.

Special characteristics Amphibious capability (max. speed 5 kph). NBC and air conditioning systems, together with a winch, are available.

Variants Version mounting 90 mm gun. Has many common automotive components with the Type 6614 APC, which is being built under licence by the Republic of Korea.

Manufacturer Fiat – OTO–Melara INVECO Defence Vehicle Division, S. Mauro Torinese.

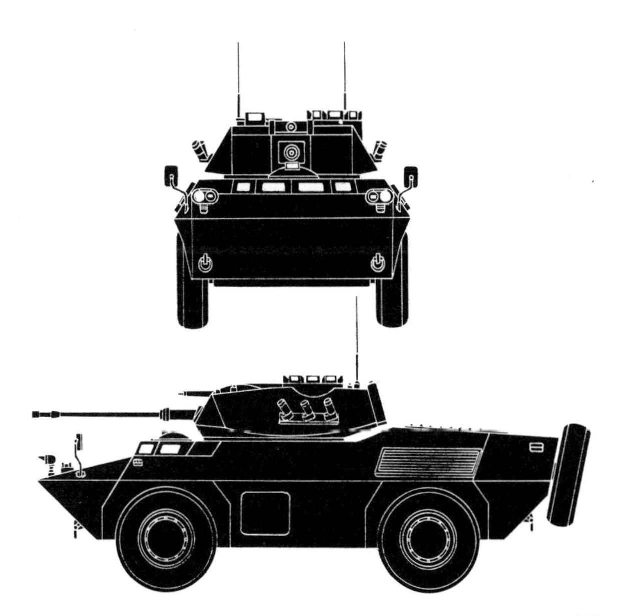

Role 155 mm self-propelled howitzer.

History A private venture by OTO–Melara, development began in 1977, with the first prototype appearing in 1981. It entered production a year later. The hull is essentially the same as the OF-40 main battle tank (see separate entry).

Employment Libya, Nigeria.

Vehicle data *Crew* 5 (commander, driver, gunner, charge handler, magazine operator), *height* 2.874 m, *width* 3.386 m, *length* 9.6 m (11.474 m with gun fully forward), *weight* 46,000 kg, *max. speed* 60 kph, *range* 400 km, *armament* 1 × 155 mm, *engine* multi-fuel developing 750 bhp at 2200 rpm.

Special characteristics Night driving aids.

Variants Argentina uses the Palmaria turret mounted on an extended TAM tank chassis.

Manufacturer OTO–Melara.

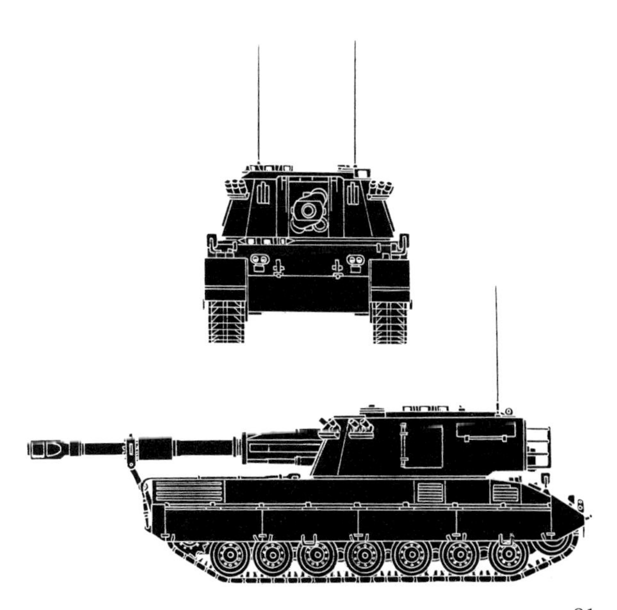

Role Main battle tank.

History The design was initiated in 1964 as a replacement for the Type 61. Prototypes were completed in 1969 and production commenced in 1975. It entered service with the Self-Defence Force in 1976.

Employment Japan.

Vehicle data *Crew* 4 (commander, gunner, driver/operator, driver), *height* 2.48 m, *width* 3.18 m, *length* 6.85 m (9.088 m with gun forward), *weight* 38,000 kg, *max. speed* 53 kph, *range* 500 km, *armament* 1 × 105 mm rifled gun (APDS, HESH, HEAT), 1 × 7.62 mm coaxial mg, 1 × 12.7 mm AA mg, *engine* Mitsubishi 10ZF Type 21 developing 750 bhp at 2200 rpm.

Special characteristics Night-fighting and -driving aids; NBC protection system; deep fording capability using a snorkel, autoloader and laser rangefinder. Hydropneumatic suspension enables vehicle height to be adjusted.

Variants Type 78 ARV.

Manufacturer Mitsubishi Heavy Industries.

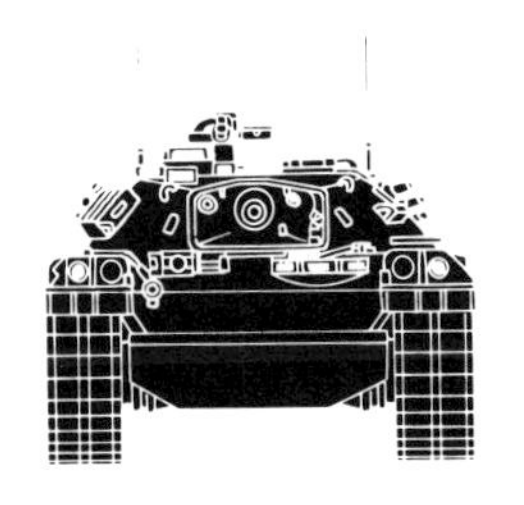

Role Armoured personnel carrier.

History Design and development of a replacement for the Type 60 APC began in the early 1960s, and Mitsubishi Heavy Industries and Komanton Steelworks each produced two prototypes in 1969. These were evaluated and the former version chosen, although it incorporated elements of the latter. Production was scheduled to begin in 1973, but budgetary problems meant that it was not commenced until 1978, and only in 1980 did it begin to enter service.

Employment Japan.

Vehicle data *Crew* 3 (commander, driver, gunner) + 12 infantrymen, *height* 1.7 m (2.2 m including 12.7 mm mg), *width* 2.8 m, *length* 5.60 m, *weight* 13,300 kg, *max. speed* 60 kph, *range* 300 km, *armament* 1 × 12.7 mm (0.50

in Browning), 1 × 7.62 mm mg, *engine* Mitsubishi 4 ZF air-cooled diesel developing 300 hp at 2200 rpm.

Special characteristics Fully amphibious using tracks, with max. speed of 7 kph. NBC protection system and night-viewing aids. Has firing ports in sides and rear of hull.

Variants Type 75 artillery observation post vehicle with ground wind measuring equipment.

Manufacturer Mitsubishi Heavy Industries.

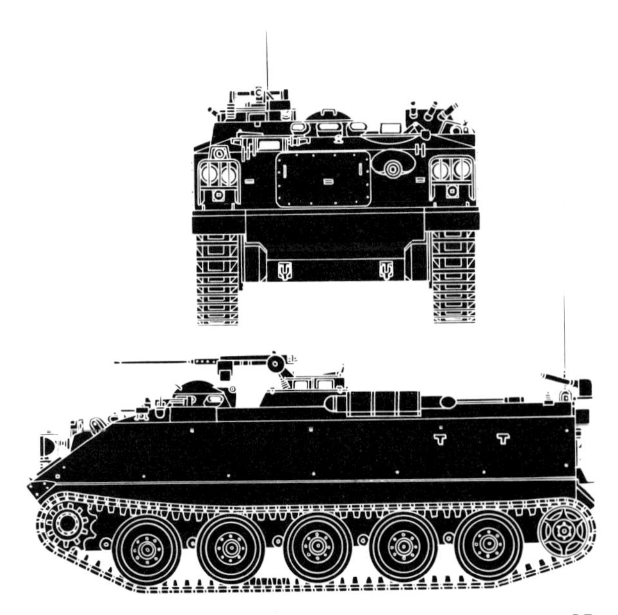

Role Command and communications vehicle.

History In 1974, in response to a Japanese Ground Self-Defence Force requirement, two light armoured vehicle designs, 4 × 4 and 6 × 6, were produced. It was decided to pursue the latter and four prototypes were produced in 1980. A production order followed and the vehicle entered service in 1984.

Employment Japan.

Vehicle data *Crew* 8 (driver, front hull gunner, commander and five), *height* 2.38 m, *width* 2.48 m, *length* 5.72 m, *weight* 13,600 kg, *max. speed* 100 kph, *range* 600 km approx., *armament* 1 × 7.62 mm pintle mounted mg (forward crew compartment), 1 × 12.7 mm pintle mounted mg (rear crew compartment), *engine* Isuzu 10PBI diesel developing 305 bhp at 2,600 rpm.

Special characteristics The design is unusual in that the engine is mounted between the forward and rear crew compartments.

Variants The Type 87 reconnaissance vehicle uses the same 6×6 chassis.

Manufacturer Mitsubishi Heavy Industries.

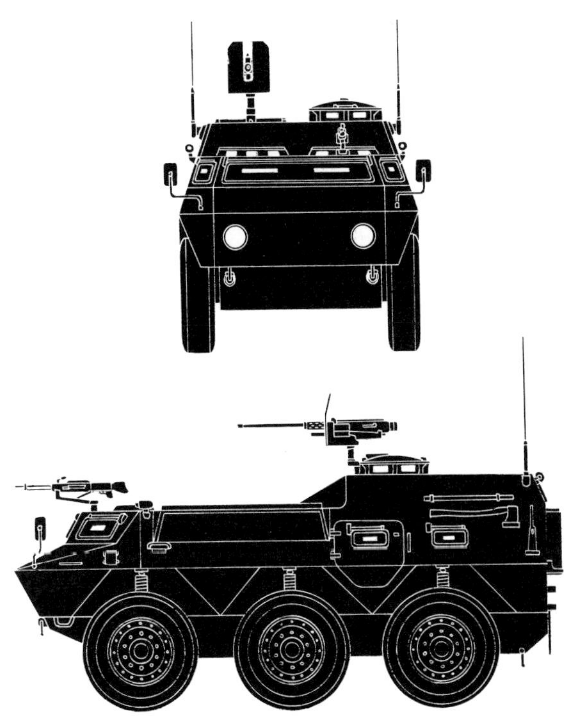

Role Armoured personnel carrier.

History Development commenced in 1956 with prototypes being produced two years later. Production ran from 1964–8. It has now been largely replaced by the PR1 (US XM765 AIFV) in the Dutch Army and is only found with Territorial Army units.

Employment Netherlands, Portugal, Surinam.

Vehicle data *Crew* 2 (commander, driver) + 10 infantrymen, *height* 2.37 m, *width* 2.40 m, *length* 6.23 m, *weight* 12,000 kg, *max. speed* 80 kph, *range* 500 km, *armament* 1 × 12.7 mm mg, *engine* DAF DS 575 diesel developing 165 bhp at 2400 rpm.

Special characteristics Night-fighting and -driving aids.

Variants PW1-S(PC) platoon commander's vehicle (additional radios); PWCO ACV; PW-GWT ambulance;

PW-V cargo carrier; PW-MT mortar tractor towing French 120 mm Brandt mortar, PWAT with TOW ATGW, PWRDR with ZB298 ground radar.

Manufacturer DAF, Eindhoven.

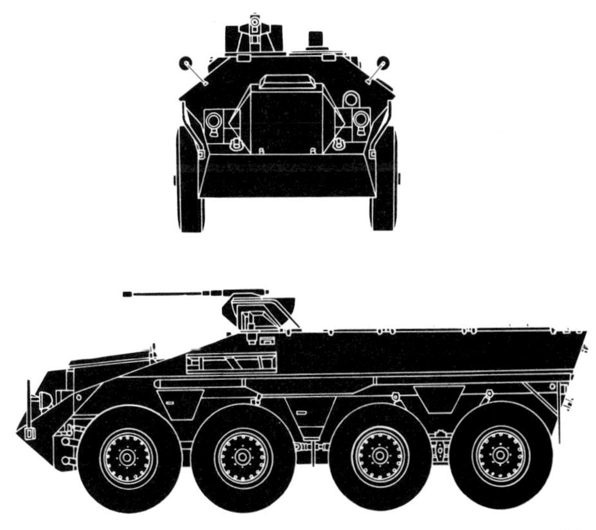

Role Armoured reconnaissance vehicle and personnel carrier.

History This range of vehicles was developed in the mid 1960s to meet a number of Portuguese military requirements, and bears a marked resemblance to the US Cadillac Gage Commando.

Employment Lebanon, Libya, Peru, Philippines, Portugal.

Vehicle data *Crew* 3–11 (commander, gunner, driver and others according to variant), *height* 2.26 m with turret, 1.84 m without turret, *width* 2.26 m, *length* 5.61 m, *weight* 7300 kg, *max. speed* 99 kph, *range* 900 km (petrol engine) 1450 km (diesel engine), *armament* variable, *engine* Model M75 V-8 petrol developing 210 bhp at 4000 rpm or V-6 diesel.

Special characteristics Fully amphibious, being

propelled by its wheels at a max. speed of 7 kph. Night-viewing equipment. All models have a winch mounted on the front.

Variants V-200 APC (turret with 2 × 7.62 mm, or 2 × 5.56 mm, or 1 × 12.7 mm and 1 × 7.62 mm – see silhouette); V-300 armoured car (turret with 20 mm or 25 mm cannon, or 35 mm gun); V-400 armoured car (with 90 mm Mecam or Bofors 90 mm gun – see photograph); V-500 command and communications (without turret); V-600 mortar carrier (81 mm or 120 mm); V-700 ATGW vehicle (HOT or Swingfire); V-800 ambulance; V-900 recovery vehicle; V-1000 internal security vehicle.

Manufacturer BRAVIA

Role Infantry fighting vehicle.

History This machine was designed for the South African Army by Sandoek-Astral, who also build the South African version of the Panhard AML (Eland), as a replacement for the ageing fleet of British Saracen APCs. It entered service in the mid 1970s.

Employment Morocco, South Africa.

Vehicle data *Crew* 3 (commander, gunner, driver) + 7 infantrymen, *height* 3.11 m, *width* 2.7 m, *length* 7.21 m, *weight* 17,000 kg, *max. speed* 105 kph, *range* 800 km (approx.), *armament* 1 × 20 mm cannon, 2 × 7.62 mm mgs (coaxial and AA), *engine* 6-cylinder turbocharged diesel.

Special characteristics 7.62 mm mg can be mounted over rear hull.

Variants Ratel 20 is the standard version. Ratel 60 has a 60 mm gun-mortar and Ratel 90 a 90 mm gun. There is also an 8 × 8 logistic version and 6 × 6 ACV.

Manufacturer Sandoek-Astral.

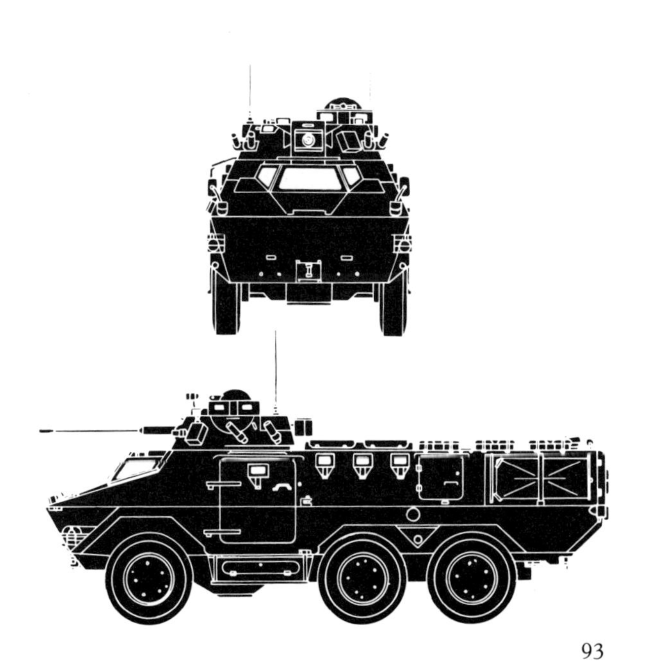

Role Reconnaissance vehicle.

History This was developed to meet a Spanish Army requirement and the prototypes were completed during 1977–78. The vehicle is now in full production.

Employment Spain.

Vehicle data *Crew* 5 (commander, gunner, driver, 2 scouts), *height* 2 m (to hull top), *width* 2.5 m, *length* 6.25 m, *weight*, 13,750 kg, *max. speed* 100 kph, *range* 800 km, *armament* 1 × 25 mm cannon, 1 × 7.72 mm coaxial mg, *engine* Pegaso model 9157/8 diesel developing 306 bhp at 2200 rpm.

Special characteristics Amphibious with a top speed of 3 kph (using its wheels) and 9 kph (hydrojets), NBC protection system and night fighting and driving devices.

Variants A number of different turrets are available ranging from 12.7 mm mg to 90 mm gun. The BMR APC series, which includes a wide range of variants, has the same 6 × 6 chassis.

Manufacturer Empresa Nacional de Autocamiones SA.

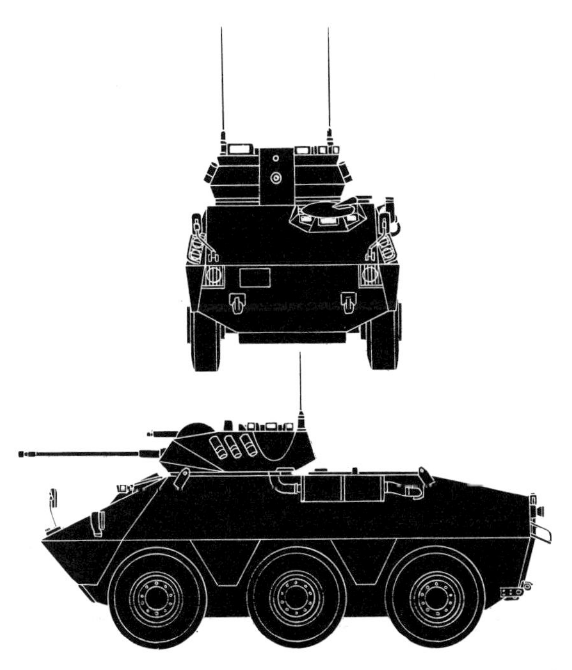

STRIDSVAGN (STRV 103B OR S TANK)

Role Main battle tank.

History The concept for this turretless tank originated in the mid 1950s. The first prototype was built in 1961 and production took place 1966–71 with some 300 being built for the Swedish Army.

Employment Sweden.

Vehicle data *Crew* 3 (commander, driver/gunner, radio operator), *height* 2.14 m, *width* 3.6 m, *length* 9.8 m (including gun fully forward), *weight* 39,000 kg, *max. speed* 50 kph, *range* 390 km, *armament* 1 × 105 mm rifled gun (APDS, HE, Smoke), 2 × 7.62 mm hull-mounted mgs, 1 × 7.62 mm AA mg, *engine* Rolls-Royce K60 diesel developing 240 bhp at 3650 rpm or Boeing gas-turbine developing 490 shp at 38,000 rpm.

Special characteristics This novel AFV has hydropneumatic suspension to enable gun to be laid on target. Autoloader; amphibious using flotation screen with max. speed of 6 kph; dozer blade mounted; night driving and -fighting aids.

Variants Original production model, Strv 103A had no flotation screen or dozer blade fitted. These were later brought up to 103B standard. Current modernisation programme includes fitting Detroit Diesel 6V-53T engine and laser rangefinder.

Manufacturer Bofors.

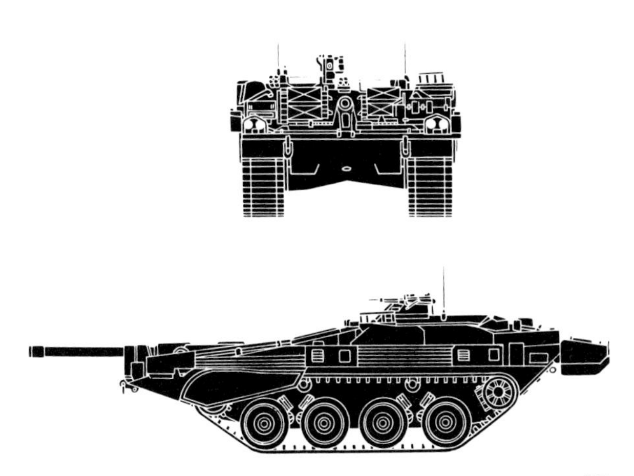

INFANTERIKANONVAGN 91 (IKV-91)

Role Light tank/tank destroyer.

History Designed to meet a Swedish Army requirement to rationalize existing concepts on light tanks and tank destroyers, the prototype was built in 1969. Production commenced in 1974, and it entered service with the Swedish Army a year later, replacing STRV 74, IKV-102 and IKV-103. Production was completed in 1977.

Employment Sweden.

Vehicle data *Crew* 4 (commander, gunner, loader, driver), *height* 2.355 m, *width* 3 m, *length* 6.41 m (8.835 m with gun fully forward), *weight* 15,500 kg, *max. speed* 69 kph, *range* 550 km, *armament* 1 × 90 mm low-pressure gun (HEATFS, HEFS), 1 × 7.62 mm coaxial mg,

1 × 7.62 mm AA mg, *engine* Volvo-Penta TD 120A diesel developing 295 bhp at 2200 rpm.

Special characteristics Amphibious (propelled by its tracks at max. speed of 7 kph); night-driving aids; NBC protection system.

Variants ATGW (HOT) launcher vehicle (Pvrbv 551) and SAM launcher carrier (Bofors RBS-70 SAM) (Lvrbv 701); IKV 91-105 with 105 mm low recoil gun.

Manufacturer Hägglunds and Soner, Örnsköldsvik.

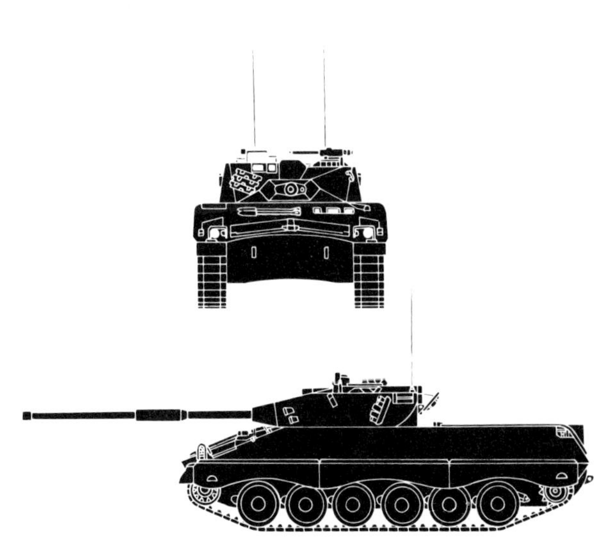

155 MM VK 155 BANDKANON 1A

Role Self-propelled artillery gun.

History Designed and developed by Bofors in the late 1950s, the prototype was completed in 1961. It entered service with the Swedish Army in 1966. It uses many components of the S-tank. Production was completed in 1968.

Employment Sweden.

Vehicle data *Crew* 6 (commander, driver, loader, layer, radio operator, AA mg gunner), *height* 3.85 m, *width* 3.37 m, *length* 11 m (including gun), *weight* 53,000 kg, *max. speed* 28 kph, *range* 230 km, *armament* 1 × 155 mm gun, 1 × 7.62 mm AA mg, *engine* Rolls-Royce K60 diesel developing 240 bhp at 3750 rpm or Boeing model 502/10MA gas turbine developing 300 shp at 38,000 rpm.

Special characteristics Night-driving aids. Automatic loader using two 7-round magazines which can achieve rate of fire of 14 rds/min.

Variants Nil.

Manufacturer Bofors.

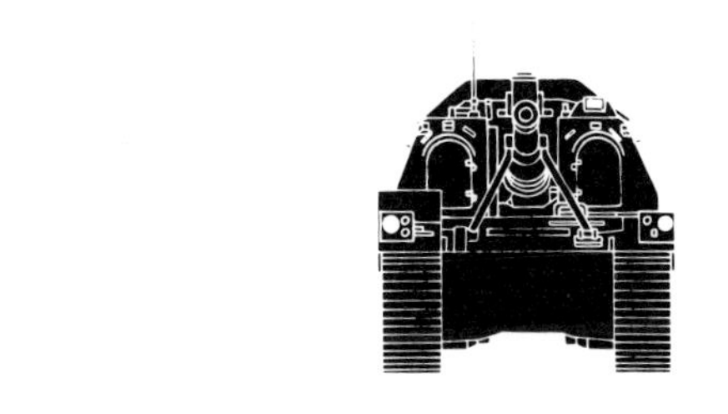

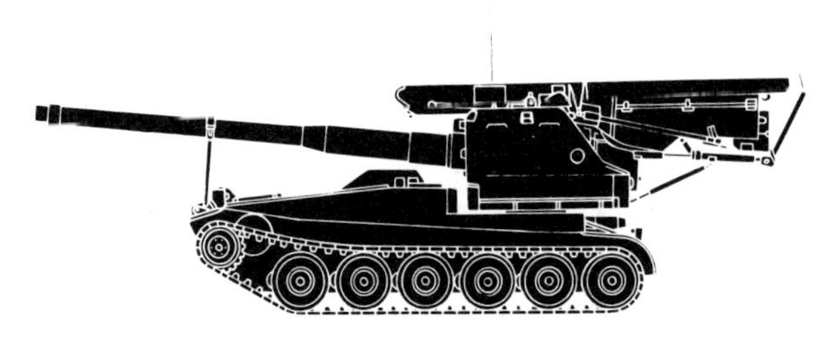

Role Main battle tanks.

History Design began in the early 1950s, and the prototype, known as Pz58, mounting a 20 pdr (84 mm), was completed in 1958. It was then decided to mount a version of the British 105 mm gun on production models, and 150 of these were produced as Pz61 over 1964–6. The improved Pz68 made its appearance in 1971, with 170 being built between then and 1973.

Employment Switzerland

Vehicle data *Crew* 4 (commander, gunner, loader, driver), *height* 2.72 m (Pz61), 2.74 m (Pz68), *width* 3.06 m (Pz61), 3.14 m (Pz68), *length* 6.78 m (Pz61), 6.9 m (Pz68), *weight* 38,000 kg (Pz61), 39,700 kg (Pz68), *max. speed* 50 kph (Pz61), 55 kph (Pz68), *range* 300 km, *armament* 1 × 105 mm rifled gun (APDS, HEAT, HEP, Canister,

Smoke), 1 × 20 mm coaxial mg (7.5 mm on Pz68), 1 × 7.5 mm AA mg, *engine* MTU MB 837 diesel developing 630 bhp at 2200 rpm (704 bhp at 2200 rpm on Pz68). Photograph and silhouettes are Pz68.

Special characteristics Night-driving aids; NBC protecting systcm.

Variants Pz68 MK2 has thermal sleeve on gun and Pz68 MK3/4 has larger turret. ARV (Entpannungspanzer 65), bridgelayer (Brückenpanzer 68) both on Pz68 chassis.

Manufacturer Federal Construction Works.

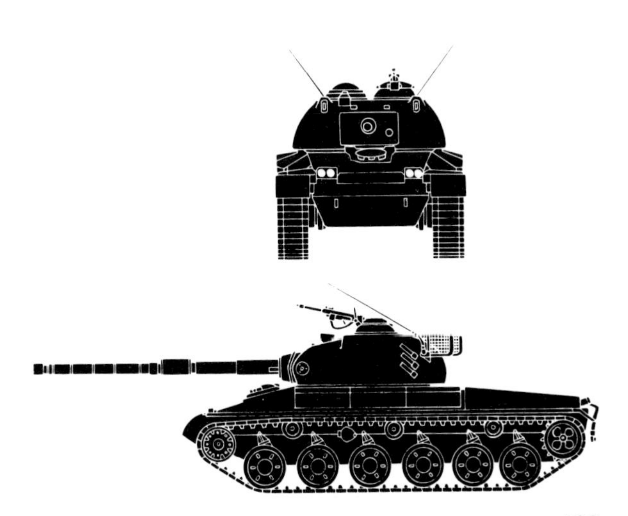

Role Multi-role light AFV family.

History MOWAG commenced work on this venture in the early 1970s using experience gained on the Puma 6 × 6 APC. First production vehicles appeared in 1976, and in 1977 General Motors of Canada was awarded a contract to build variants of the 6 × 6 version under licence for the Canadian Army. Further export orders followed.

Employment Canada, Chile, Ghana, Liberia, Nigeria, Peru, Switzerland, USA.

Vehicle data (6 × 6) *Crew* 3–12, *height* 1.85 m (hull top), *width* 2.5 m, *length* 5.84 m, *weight* 9600 kg, *max. speed* 100 kph, *range* 600 km, *armament* variable, *engine* (Canadian) Detroit diesel 6V53T turbocharged developing 275 bhp at 2800 rpm.

Special characteristics All models are fully amphibious using two propellers mounted in rear with max. speed of 10 kph. NBC protection system and night-viewing and -fighting aids.

Variants 4 × 4 (known as Spy) has either 12.7 mm turret or twin 12.7 mm or 7.62 mm turret. 6 × 6 standard model is an MICV with weapons fits ranging through 20 mm cannon to 90 mm anti-tank and 120 mm mortar. 8 × 8 variants (Piranha) also exist. Canadian versions are Cougar (wheeled fire-support vehicle with 76 mm gun in Scorpion turret – see photograph), Grizzly (APC with 12.7 mm mg – see silhouettes) and Husky (wheeled maintenance and recovery vehicle). Version with Cockerill 90 mm gun is manufactured under licence in Chile. 8 × 8 (known as Shark) has 20, 25 or 30 mm cannon.

Manufacturer MOWAG Motorwagenfabrik and Diesel Division of General Motors of Canada.

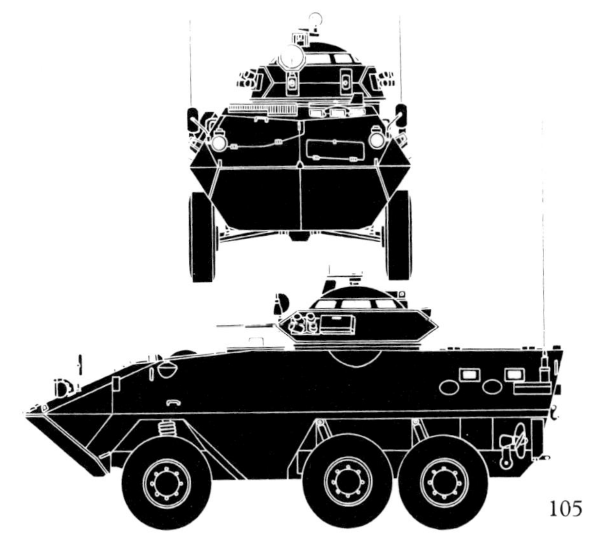

Role Main battle tank.

History Development commenced in 1944 and 6 prototypes were sent to Germany in May 1945, arriving just too late to be tested in combat. It was originally armed with a 17 pdr (76 mm), but this was replaced by a 20 pdr (84 mm) with the introduction of the Mk3 in the early 1950s, and finally by a 105 mm in the early 1960s. It is virtually obsolete in the British Army. Production was completed in 1961.

Employment Austria (turrets only), Denmark, Israel, Jordan, Netherlands, Kuwait, Singapore, Somalia, South Africa, Sweden, Switzerland, United Kingdom (105 mm AVRE).

Vehicle data (Mk 13). *Crew* 4 (commander, gunner, loader/operator, driver), *height* 3.009 m, *width* 3.39 m, *length* 7.823 m (9.854 m with gun over front), *weight* 51,820 kg, *max. speed* 34.6 kph, *range* 190 km, *armament* 1 × 105 mm rifled gun (APDS, APDSFS, HESH, HEAT, Smoke, Canister), 1 × 7.62 mm coaxial mg,

1 × 12.7 mm rmg, 1 × 7.62 mm AA mg, *engine* Rolls-Royce Meteor Mk4B petrol developing 650 bhp at 2550 rpm.

Special characteristics Night-fighting and -driving aids; dozer blade can be fitted.

Variants Mk1–Mk13 with sub-marks; Mk5 bridgelayer; Mk5 AVRE (with 165 mm demolition charge projector); Mk2 ARV; Mk5 ARK bridgelayer; BARV; AVRE with 105 mm gun, mine plough and 5-man crew (co-driver); Vickers have produced a modified version with improved automotive and target acquisition features, and the Israelis use a diesel engine, as does S. Africa (Olifant), Swedish Strv 101/102 upgrade to Strv 104 with IFCS and diesel engine. AA version with Marksman (2 × 35 mm) turret.

Manufacturer Royal Ordnance Factory, Leeds; Vickers; Leyland Motors.

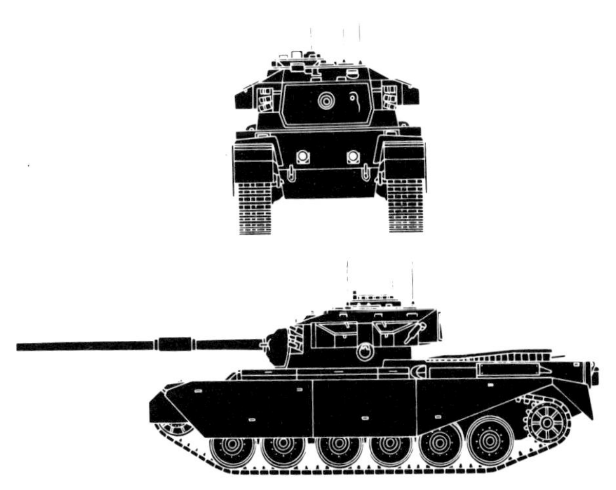

Role Main battle tank.

History Designed as the replacement for both Centurion and Conqueror (heavy tank with 120 mm gun produced in mid 1950s to counter the Russian JS-3). The prototype was completed in 1959, and the tank entered service with the British Army in 1967.

Employment Iran (Shir), Iraq, Jordan (Khalid), Kuwait, Oman, United Kingdom.

Vehicle data (Mk5) *Crew* 4 (commander, gunner, loader/operator, driver), *height* 2.895 m, *width* 3.657 m (including searchlight), *length* 7.52 m (10.79 m with gun forward), *weight* 55,000 kg, *max. speed* 48 kph, *range* 500 km, *armament* 1 × 120 mm rifled gun (APDSFS, APDS, HESH, Smoke, Canister with separated ammunition), 1 × 12.7 mm rmg, 1 × 7.62 mm coaxial mg,

1 × 7.62 mm AA mg, *engine* Leyland L60 No 4 Mk8A diesel developing 750 bhp at 2100 rpm.

Special characteristics Deep fording capability using snorkel exists, although not used by British Army; night-fighting and -driving aids; NBC protection system. Improved fire-control system (IFCS), incorporating laser rangefinder and ballistic computer, is being introduced into service.

Variants Mks range from 1 to 12 incorporating various improvements, mainly automotive and target acquisition aspects. Mks 9–12 are earlier Mks with IFCS. ARV and bridgelayer versions exist and ARRV (to support Challenger). A version with improved armour and IFCS was produced for Iran (Shir II) but not delivered, but has been further developed for the British Army as Challenger. Programme of installing 800 bhp diesel and TN12 transmission on in-service tanks is underway. Chieftain 900 with Chobham armour and further uprated engine and Chieftain-Sabre (2 × 30 mm AA turret) also exist.

Manufacturer Vickers; Royal Ordnance Factory, Leeds.

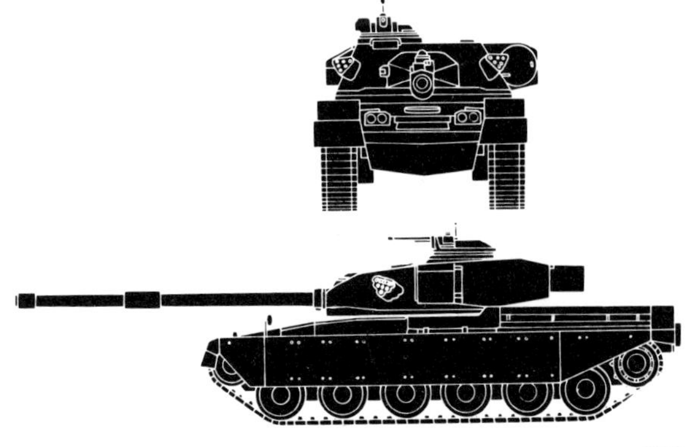

Role Main battle tank.

History In the early 1970s the UK and FRG combined to develop a main battle tank for the late 1980s. Differences in design philosophy led to cancellation in 1979, and Britain embarked on a unilateral project (MBT 80). At the same time, Iran had ordered an advanced version of Chieftain, Shir II, but this was cancelled by the Revolution. Budgetary problems resulted in a decision to adopt Shir II instead of MBT 80, and this will replace half the British Army Chieftain fleet. It began to enter service in 1983.

Employment United Kingdom.

Vehicle data *Crew* 4 (commander, gunner, loader,

driver), *height* 2.89 m, *width* 3.52 m, *length* 9.86 m (11.55 m with gun fully forward), *weight* 62,000 kg, *max. speed* 60 kph, *range* 350 km, *armament* 1 × 120 mm rifled gun (APDS, APDSFS, HESH, Smoke, Canister) 2 × 7.62 mm mg (coaxial and AA), *engine* Rolls-Royce CV12 TVA diesel developing 1200 bhp at 2300 rpm.

Special characteristics NBC protection system, night-fighting and -driving aids, Chobham armour, integrated fire-control system (IFCS) with laser rangefinder, separated main armament ammunition.

Variants Challenger Mk2 with Thermal Observation and Gunnery System (TOGS) and Challenger ARRV (under development).

Manufacturer Royal Ordnance Factory, Leeds.

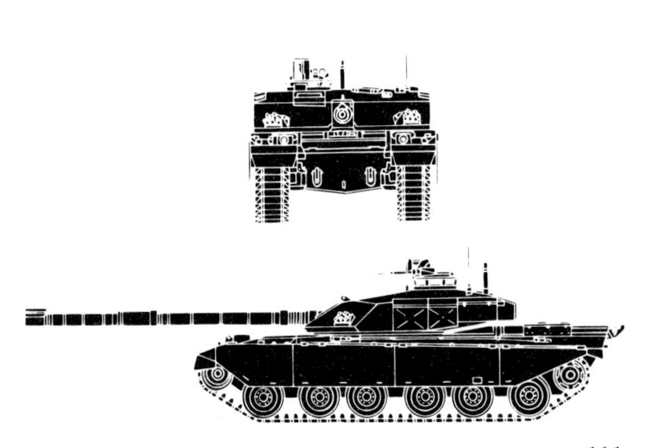

VICKERS MBT (VIJAYANTA)

Role Main battle tank.

History It originated from a private venture by Vickers for a light tank armed with a 20 pdr gun and Vigilant ATGW. In 1960 India expressed an interest in the scheme, and the design was adapted to mount a 105 mm gun, as well as incorporating fire-control equipment similar to Chieftain. The first production tank was delivered to India in 1965, and Indian-built production models first appeared in 1969.

Employment India (Vijayanta), Kenya, Kuwait, Nigeria.

Vehicle data *Crew* 4 (commander, gunner, loader/operator, driver), *height* 2.64 m, *width* 3.168 m, *length* 7.92 m (9.728 m with gun forward), *weight* 38,600 kg, *max. speed* 56 kph, *range* 480 km, *armament* 1 × 105 mm rifled gun (APDS, HESH, Smoke), 1 × 12.7 mm rmg, 1 × 7.62 mm coaxial mg, *engine* Leyland L60 Mk4B developing 650 bhp at 2670 rpm.

Special characteristics Amphibious using flotation screen and track propulsion (max. speed 6.4 kph); NBC protection system, night-driving and -fighting equipment.

Variants Mk 2 incorporating Swingfire ATGW was prototype only. Mk 3 with Chieftain-type turret was introduced (see silhouettes and photograph) in 1973. ARV and bridgelayer variants are also in production. Valiant, a variant with 120 mm gun, Challenger-like turret and Chobham armour has been introduced. India has used Vickers chassis for 130 mm SP gun. Vickers also have a 155 mm SP gun and AA (Marksman 2 × 35 mm turret) versions.

Manufacturer Vickers; Avadi, Madras, India.

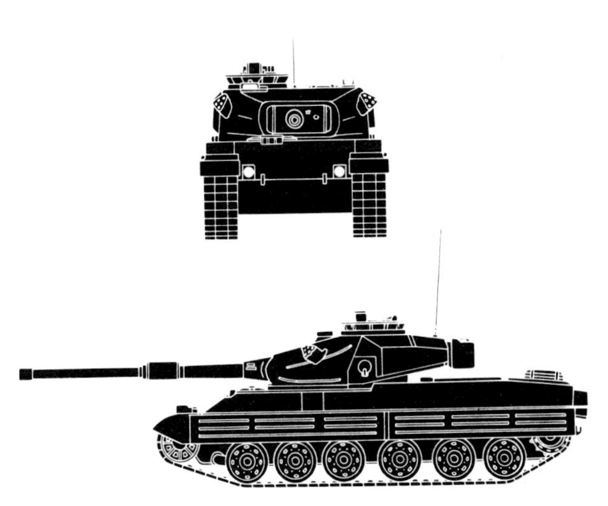

Role Scout car.

History Design commenced shortly after the end of World War II. The prototype, a Mk 1 without a turret, was completed in 1949, and production on the Mk 1 and Mk 2 began in 1952, continuing until 1971. In the British Army Ferret Mk 2 has now been replaced by Fox in reconnaissance units. Production was completed in 1971.

Employment Bahrain, Burkina Faso, Burma, Cameroon, Central African Republic, Gambia, Ghana, Indonesia, Jordan, Kuwait, Libya, Madagascar, Malaysia, New Zealand, Nigeria, Portugal, Qatar, South Africa, Sri Lanka, Sudan, United Arab Emirates, United Kingdom, Yemen, Zimbabwe.

Vehicle data (Mk 2/3) *Crew* 2 (commander, driver), *height* 1.879 m, *width* 1.905 m, *length* 3.835 m, *weight* 4400 kg, *max. speed* 90 kph, *range* 300 km, *armament* 1 × 7.62 mm mg, *engine* Rolls-Royce B60 Mk6A petrol developing 129 bhp at 3750 rpm.

Special characteristics Deep fording capability using special kit to a depth of 1.524 m.

Variants Mk1 liaison (without turret); Mk2/6 ATGW (Vigilant missile launcher on either side of turret); Mk3 amphibious (Mk1 with flotation screen and big wheels – propelled by wheels at max. speed of 3.8 kph); Mk4 (Mk3 with turret); Mk5 ATGW (modified turret with twin Swingfire missile launchers on either side). In prototype is Ferret 80 with 2/3 man crew, diesel engine, 7.62 mm mg/25 mm Chain Gun/TOW ATGW and redesigned body and turret.

Manufacturer Daimler Company.

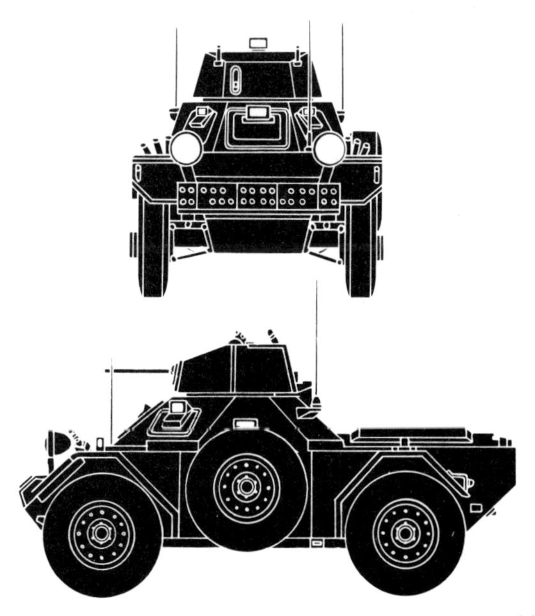

Role Armoured car.

History Saladin was developed as a result of the experience gained in armoured cars during World War II. Prototypes first appeared in 1953, and it entered production in 1958. It has been superseded by Scorpion in the British Army. Production was completed in 1972.

Employment Bahrain, Ghana, Honduras, Indonesia, Jordan, Kenya, Kuwait, Lebanon, Libya, Nigeria, Portugal, Qatar (Saracen only), Sierra Leone, Sri Lanka, Sudan, Tunisia, Uganda (Saracen only), United Arab Emirates, Yemen.

Vehicle data *Crew* 3 (commander, gunner, driver), *height* 2.93 m, *width* 2.54 m, *length* 4.93 m (5.284 m with gun fully forward), *weight* 11,590 kg, *max. speed* 72 kph, *range* 400 km, *armament* 1 × 76 mm rifled gun (HESH, HE, Smoke), 2 × 7.62 mm mgs (commander's, coaxial), *engine* Rolls-Royce B80 Mk6A petrol developing 160 bhp at 3750 rpm.

Special characteristics Nil.

Variants Saracen APC. A version mounting Swingfire ATGW reached prototype stage.

Manufacturer Alvis.

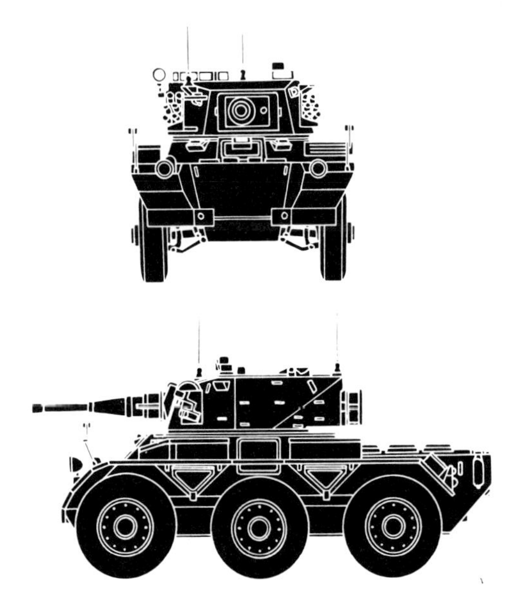

Role Reconnaissance vehicle.

History In the early 1960s work started on an armoured vehicle reconnaissance (AVR) to replace Saladin and Ferret. The original intention was for it to combine its reconnaissance role with ATGW and air-portability capabilities. This proved impracticable, and the result was the CVR(T) family and CVR(W) Fox. The Scorpion prototype was completed in 1969, and it entered production in 1972, followed successively by the other 6 members of the family. Belgium has a 20 percent stake in the project.

Employment Belgium, Brunei, Honduras, Iran, Ireland, Kuwait, Malaysia, New Zealand, Nigeria, Oman, Philippines, Singapore, Spain, Tanzania, Thailand, United Arab Emirates, United Kingdom.

Vehicle data *Crew* 3 (commander, gunner, driver),

height 2.096 m, *width* 2.184 m, *length* 4.388 m, *weight* 7960 kg, *max. speed* 87 kph, *range* 650 km, *armament* 1 × 76 mm rifled gun (HESH, HE, Smoke, Canister), 1 × 7.62 mm coaxial mg, *engine* Jaguar ohc 4200 cc, petrol developing 195 bhp at 4750 rpm.

Special characteristics Night-fighting and -driving aids; amphibious using a flotation screen with max. speed of 6.44 kph; NBC protection system.

Variants Scimitar (mounting 30 mm Rarden gun). Scorpion 90 with Cockerill 90 mm low-pressure gun (Malaysia). Scorpion turrets are to be found on a number of AFV's, including Canadian Cougar and Australian M113. Diesel version exists, and laser rangefinder and night vision improvements are available.

Manufacturer Alvis; Malines Plant, Belgium.

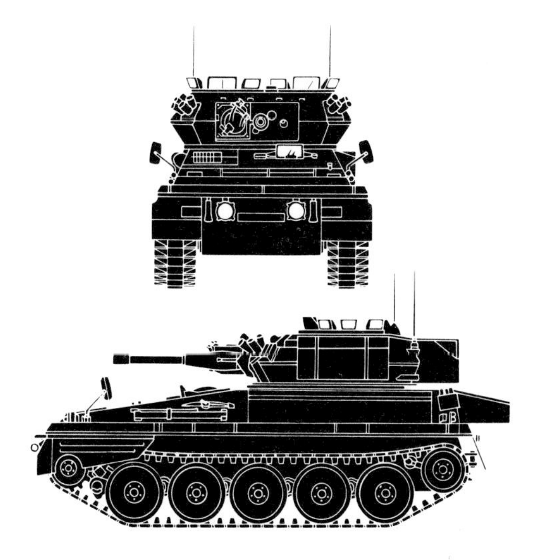

Role Reconnaissance vehicle.

History Fox was developed over the period 1965–70 as a successor to Ferret in the reconnaissance role. It entered service with the British Army in 1975. A turretless version to replace Ferret Mk1 in the liaison role, known as Vixen, was also developed but never entered production.

Employment Malawi, Nigeria, United Kingdom.

Vehicle data *Crew* 3 (commander, gunner, driver), *height* 2.03 m, *width* 2.134 m, *length* 4.166 m, *weight* 6120 kg, *max. speed* 105 kph, *range* 430 km, *armament* 1 × 30 mm Rarden gun, 1 × 7.62 mm coaxial mg, *engine*

Jaguar ohc 4200 cc petrol developing 190 bhp at 4500 rpm.

Special characteristics Amphibious using flotation screen with max. speed of 5 kph; night-fighting and -driving aids.

Variants Panga (with one-man turret mounting 7.62 mm or 12.7 mm mg), Fox-Milan (with 2 Milan ATGW launchers, 7.62 mm Hughes chain gun or 7.62 mm mg).

Manufacturer Royal Ordnance Factory, Leeds, with turrets subcontracted to Alvis.

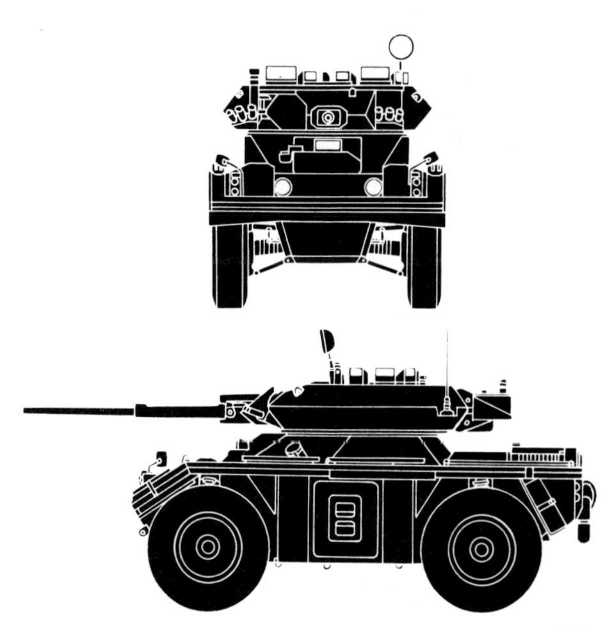

Role 105 mm self-propelled gun.

History Design began in the late 1950s, with the prototype appearing in 1961. Production took place 1964–7. It uses the FV 432 chassis. It was replaced in the British Army by the now cancelled FRG/Italian/UK SP70 155 mm SP howitzer.

Employment India, United Kingdom.

Vehicle data *Crew* 4 (commander, driver, layer, loader), *height* 2.489 m, *width* 2.641 m, *length* 5.84 m, *weight* 16,556 kg, *max. speed* 48 kph, *range* 390 km, *armament* 1 × 105 mm gun, 1 × 7.62 mm mg, *engine* Rolls-Royce K60 Mk4G multi-fuel developing 240 bhp at 3750 rpm.

Special characteristics NBC protective system;

night-driving aids; amphibious (using tracks and flotation screen with max. speed of 5 kph).

Variants Falcon (with 2 × 30 mm AA) appeared in prototype only. Value engineered Abbot 50 has non-essential items (NBC system, flotation screen) removed and is used by Indian Army.

Manufacturer Vickers.

Role Armoured personnel carrier.

History The prototype appeared in 1961, and production took place in 1963–71. It was originally called Trojan, but this name was later discarded because of possible confusion with a car firm of the same name. Some will be replaced by AT105 in British Army. Abbot SP 105 mm gun uses many of the automotive components.

Employment United Kingdom.

Vehicle data *Crew* 2 (commander, driver) + 10 infantrymen, *height* 2.286 m, *width* 2.8 m, *length* 5.25 m, *weight* 15,280 kg, *max. speed* 52 kph, *range* 580 km, *armament* 1 × 7.62 mm mg (some have turret-mounted 7.62 mm), *engine* Rolls-Royce K60 No 4 Mk4F developing 240 bhp at 3750 rpm.

Special characteristics Amphibious (using track propulsion and flotation screen with max. speed of 6.6 kph); NBC protection system; night-driving aids.

Variants FV434 (repair vehicle); ambulance; ACV; 81 mm mortar carrier; Wombat carrier (120 mm recoilless AT gun); surveillance vehicle (with ZB 298 radar); FV432 with Cymbeline (mortar-locating radar); FV439 communications vehicle; FV432 with FACE (artillery fire control computer); FV432 with Bar minelayer and/or Ranger mine projector; FV436 with Ptarmigan trunk communications system; FV432 with 7.62 mm mg turret; FV432 with Rarden 30 mm gun in Scimitar turret.

Manufacturer GKN Sankey.

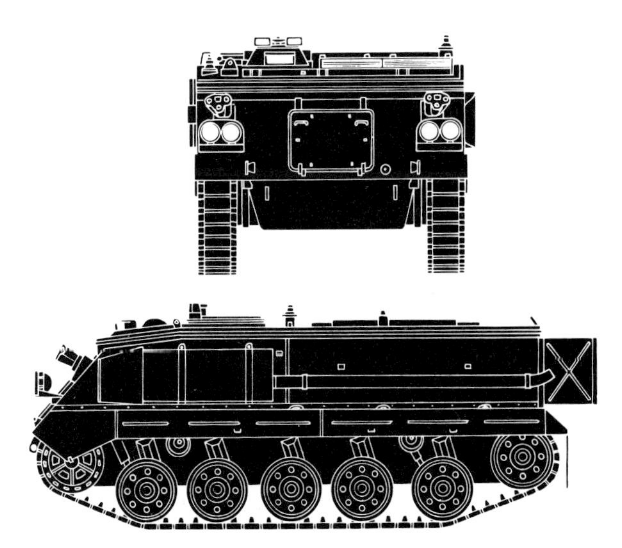

Role Armoured personnel carrier and internal security vehicle.

History Origins lie in AT 100 IS vehicle developed by GKN Sankey in 1971. This did not go into production and was superseded by AT 104. AT 105 was introduced in 1974, and took over from AT 104. It uses many Bedford commercial automotive components. In 1983 the British Army decided to buy it to replace its wheeled Saracens and Humber 1-ton APCs still in service, as well as some of its FV432s.

Employment Bahrain, Kuwait, Malaysia, Nigeria, Oman, United Kingdom.

Vehicle data (Standard 105 P model) *Crew* 2 (commander, driver) + 8 infantrymen, *height* 2.63 m, *width* 2.49 m, *length* 5.17 m, *weight* 10,670 kg, *max. speed* 96 kph, *range* 510 km, *armament* 1 × 7.62 mm mg in fixed cupola, *engine* Bedford 6-cylinder diesel developing 164 bhp at 2800 rpm.

Special characteristics Firing ports in hull sides and rear; air-conditioning system, winch barricade remover, searchlight, grenade launchers can be provided.

Variants AT 105E (with one or two 7.62 mm mgs in turret), AT 105MR (with 81 mm mortar), AT 105C ACV, AT 105A ambulance. A development of Saxon is the Simba range of AFVs with more 'workmanlike' hull and armament. A further version has the French MCT Milan ATGW turret.

Manufacturer GKN Sankey.

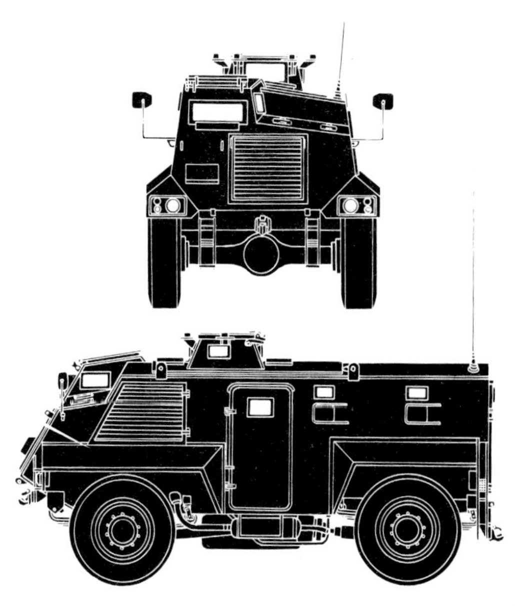

Role Mechanized infantry combat vehicle.

History Work on MCV-80 as a replacement for FV432 began in the early 1970s. GKN Sankey was awarded a production contract by the British Army in 1980, and the vehicle is expected to come into service in 1988. Economic problems have meant that not all the FV432 fleet can be replaced by it, and hence some will 'run on', while others will be replaced by AT 105 Saxon.

Employment United Kingdom.

Vehicle data *Crew* 2 (commander, driver) + 8 infantrymen, *height* 2.82 m, *width* 2.80 m, *length* 5.42 m, *weight* 20,000 kg, *max. speed* 75 kph, *range* 500 km, *armament* 1 × 30 mm Rarden gun, 1 × 7.62 mm coaxial mg, *engine* Rolls-Royce CV8 TCE V-8 diesel developing 800 bhp at 2300 rpm.

Special characteristics NBC protection system, night-driving aids. In line with British philosophy, there is no provision for firing ports in the hull.

Variants The following are planned: platoon vehicle with 7.62 mm turret; ACV (also with 7.62 mm turret); artillery command post; AEV (with EMI Ranger anti-personnel minelaying system); mortar carrier; ARRV. However, budgetary restrictions are likely to result in some of these not going into production for the British Army. Versions with various ATGW and SAM types and with 90 mm or 105 mm guns are under consideration.

Manufacturer GKN Sankey.

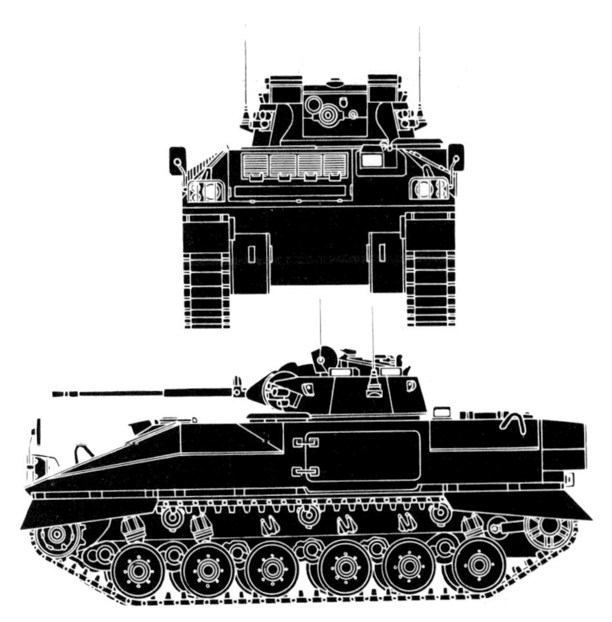

Role Internal security patrol vehicle.

History Developed in the 1960s by Short Brothers to meet a requirement of the Royal Ulster Constabulary (RUC), they were handed over to the Ulster Defence Regiment (UDR) on its formation in 1970. The vehicle consists essentially of a Ferret Mk2 turret on a Land-Rover long wheel-base chassis.

Employment Argentina, Brunei, Gulf states, Kenya, Libya, Malaysia, Thailand, United Kingdom (UDR).

Vehicle data (Mk3) *Crew* 3 (commander, gunner, driver), *height* 2.286 m, *width* 1.778 m, *length* 4.597 m, *weight* 3.360 kg, *max. speed* 88 kph, *range* 260 km (standard fuel tank) 510 km (long-range tank), *armament* 1 × 7.62 mm mg, *engine* Rover petrol developing 91 bhp at 4500 rpm.

Special characteristics Tear-gas discharger can be fitted in place of mg.

Variants Mk1 had engine developing 67 bhp at 4000 rpm, Mk2 develops 77 bhp at 4000 rpm, and both have thinner armour than Mk 3. Mk4 has 3.5 litre petrol engine and further improved armour. Other versions are Short SB501 with fully enclosed all welded armour box body and no turret. Portuguese Commando III APC is similar in shape but with different chassis (longer wheel-base) and is heavier.

Manufacturer Short Brothers and Harland.

Role Armoured personnel carrier, engineer reconnaissance vehicle and surveillance vehicle.

History Developed as part of the CVR(T) family (see under Scorpion); prototypes were produced in 1973 and it entered service with the British Army in 1977.

Employment Belgium, Brunei, Iran, Eire, United Kingdom.

Vehicle data *Crew* 3 (commander, driver, gunner) + 4 infantrymen, *height* 2.25 m, *width* 2.184 m, *length* 4.839 m, *weight* 8170 kg, *max. speed* 87 kph, *range* 640 km, *armament* 1 × 7.62 mm mg, *engine* Jaguar ohc 4200 cc petrol developing 195 bhp at 4750 rpm.

Special characteristics Can mount ZB298 surveillance radar; night-driving aids; NBC protection system.

Variants Striker (ATGW vehicle mounting 5 Swingfire launchers); Samson (ARV); Sultan (ACV); Samaritan (ambulance). The latter two have a higher profile. Stormer, with additional set of road wheels, has been developed by Alvis primarily as an MICV and can be armed with a range of weapons, from 7.62 mm mg to 20/30 mm cannon, or 76/90 mm gun. Spartan variants are Milan compact turret version and Javelin SAM version.

Manufacturer Alvis.

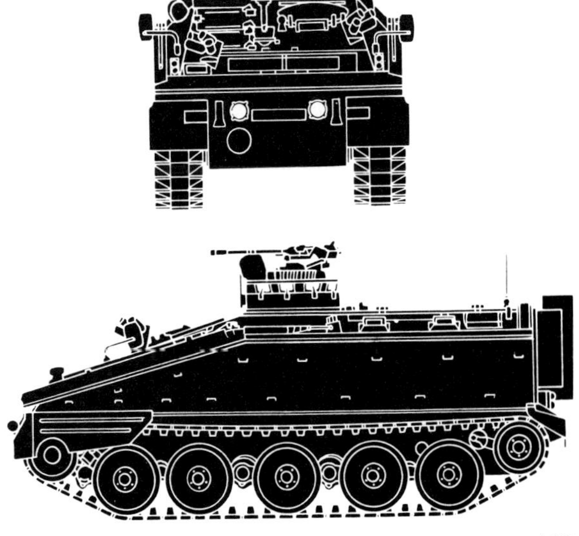

Role Main battle tank.

History Development on a successor to M47 began in 1950, and production of M48 and slightly improved M48A1 took place 1952–6. In 1954 work began on M48A2, with radical improvements to fire-control and automotive aspects, including much increased range. Production was entered two years later on this, and the diesel version M48A3 introduced shortly afterwards. During 1975–8 a further version, M48A5 (M48A3 upgunned to 105 mm), was introduced.

Employment China (Taiwan), Federal Republic of Germany (with 105 mm), Greece, Israel (with 105 mm), Iran, Jordan, Republic of Korea (South), Lebanon, Morocco, Norway, Pakistan, Portugal, Somalia, Spain, Thailand, Tunisia, Turkey (with 105 mm), USA, Vietnam.

Vehicle data (M48A3) *Crew* 4 (commander, gunner,

loader, operator), *height* 3.124 m, *width* 3.631 m, *length* 6.882 m (7.442 m with gun fully forward), *weight* 47,200 kg, *max. speed* 48 kph, *range* 460 km, *armament* 1 × 90 mm rifled gun (APDS, HEP, HEAT), 1 × 7.62 mm coaxial mg, 1 × 12.7 mm AA mg, *engine* Continental 12-cylinder AVDS-1790-2A diesel developing 750 bhp at 2400 rpm.

Special characteristics (M48A3) Night-fighting and -driving aids; NBC protection system; can deep-ford to depth of 2.438 m using snorkel; dozer blade can be fitted.

Variants M67 flamethrower; M48 AVLB. Most countries are now bringing their M48s up to A5 standard.

Manufacturer Fisher Body Division of General Motors Corporation; Ford Motor Company; Chrysler Corporation.

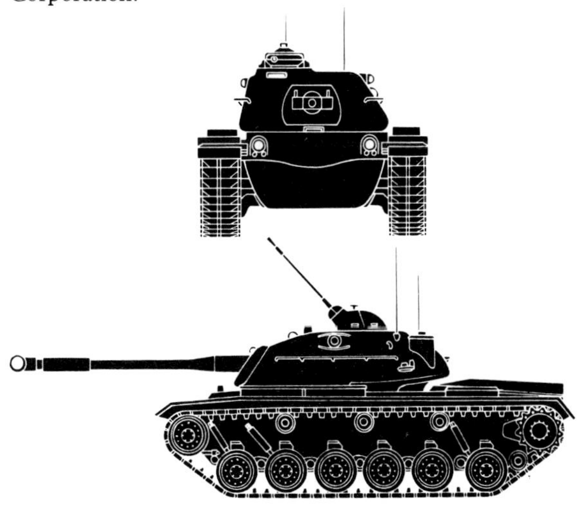

Role Main battle tank.

History Development of a successor to M48 began in 1956, with prototypes being produced in 1958. M60 entered service in 1960, and M60A1, with a modified commander's cupola, was introduced in 1962. M60A1 has recently undergone improvements in target-acquisition and automotive features, and the new version is known as M60A3. Production was completed in 1982. It is being largely replaced in US Army by M1 Abrams.

Employment Austria, Bahrain, Egypt, Iran, Israel, Italy, Jordan, Republic of Korea (South), Lebanon, Morocco, Oman, Saudi Arabia, Singapore, Somalia, Spain (AVLB), Sudan, Tunisia, Turkey, USA

Vehicle data *Crew* 4 (commander, gunner, loader, driver), *height* 3.257 m, *width* 3.631 m, *length* 6.946 m (9.309 m with gun fully forward), *weight* 49,000 kg, *max. speed* 48 kph, *range* 500 km, *armament* 1 × 105 mm rifled gun (APDS, HESH, HEAT, Smoke, Canister),

1 × 7.62 mm coaxial mg, 12.7 mm commander's mg, *engine* Continental 12-cylinder diesel AVDS-1790-2A developing 750 bph at 2400 rpm.

Special characteristics Night-fighting and -driving aids; can deep-ford using snorkel to depth of 4.114 m; NBC protection system.

Variants AVLB; M728 CEV with 165 mm demolition gun. Mineroller and dozer kits are also available. Teledyne Continental Motors have produced high-performance version with hydropneumatic suspension and appliqué armour.

Manufacturer Chrysler Corporation, Detroit Arsenal Tank Plant.

Role Main battle tank.

History Work was started on an M60 replacement in 1963, but the first concept, MBT70, a joint project with the FRG, was found to be too expensive. Work then started on XM803, but this was also found to be too complex and costly, and was cancelled by the US Congress in 1971, and the XM1 programme was instituted instead. Prototypes appeared in 1976, production began in 1980, and the M1 entered service at the end of that year.

Employment USA.

Vehicle data *Crew* 4 (commander, gunner, loader, driver), *height* 2.90 m, *width* 3.66 m, *length* 7.92 m, (9.77 m with gun fully forward), *weight* 54,430 kg, *max. speed*

72 kph, *range* 450 km, *armament* 1 × 105 mm rifled gun (XM735 improved APDSFS, HEAT, HESH, Smoke, Canister), 1 × 12.7 mm AA and 2 × 7.62 mm mgs (coaxial and AA), *engine* Lycoming ACT-1500C gas turbine developing 1500 bhp at 3000 rpm.

Special characteristics Chobham-type armour, integrated fire-control system with laser rangefinder, night-viewing and -fighting aids, NBC protection system.

Variants M1A1 with 120 mm Rheinmetall smooth-bore gun, further armour and NBC improvements is entering service. AVLB and dozer and mine-roller kits are under development.

Manufacturer General Dynamics Land Systems and Detroit Arsenal Tank Plant.

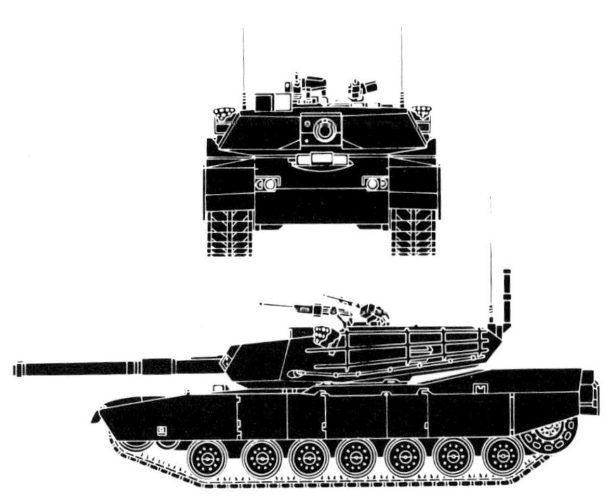

LYNX (M113C&R)

Role Reconnaissance vehicle and ACV.

History Developed by FMC as a private venture in the early 1960s, it was not taken up by the US Army, who opted for M114. Majority of components are the same as in M113A1.

Employment Canada, Netherlands.

Vehicle data (Canadian version) *Crew* 3 (commander, gunner, driver), *height* 2.171 m, *width* 2.413 m, *length* 4.597 m, *weight* 8775 kg, *max. speed* 70 kph, *range* 520 km, *armament* 1 × 12.7 mm mg, 1 × 7.62 mm mg, *engine* Detroit diesel (GMC) 6V53 developing 215 bhp at 2800 rpm.

Special characteristics Amphibious, being propelled by its tracks at max. speed of 5.6 kph; night-driving aids.

Variants Dutch basic version has side-by-side cupolas for driver and gunner, rather than gunner's cupola being to the rear as on Canadian version. Dutch also have a version with the turret modified to take an externally mounted 25 mm cannon (see photograph). Other weapons fits, including 106 mm recoilless rifle and ATGW, are available.

Manufacturer FMC Corporation.

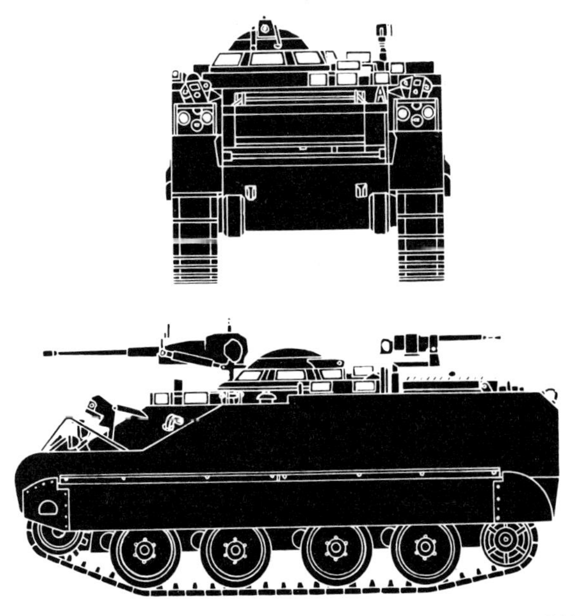

Role 175 mm self-propelled artillery gun.

History Work started in 1956 on a new family of air-portable SP artillery weapons. These were to become M107, M110 and M578 ARV. The prototype M107 was completed in 1958, using a petrol engine; it was later converted to diesel. Production commenced in late 1961, and it entered service with the US Army two years later, but has now been superseded by M110, by means of merely replacing the gun and mounting.

Employment Federal Republic of Germany, Greece, Iran, Israel, Italy, Republic of Korea (South), Spain, Turkey, United Kingdom.

Vehicle data *Crew* 5 (commander, driver, and 3 gun crew) + 8 additional gun crew carried on another vehicle, *height* 2.809 m, *width* 3.149 m, *length* 11.256 m (with gun fully forward), *weight* 28,200 kg, *max. speed* 56 kph, *range* 725 km, *armament* 1 × 175 mm, *engine* 8V71T Detroit diesel developing 405 bhp at 2300 rpm.

Special characteristics Night-driving aids.

Variants M107E1 (with automotive and gun-loading improvements). M578 ARV and M110 use same chassis.

Manufacturer FMC Corporation; Pacific Car and Foundry Company; Bowen–McLaughlin–York.

Role 155 mm self-propelled howitzer.

History Development began in 1952; the original intention was to mount a 156 mm gun, but this was altered to 155 mm in 1956. The prototype was produced in 1959, with a petrol engine, but a further prototype with a diesel engine was produced a year later. Production commenced in 1962, and it entered service with the US Army in 1964.

Employment Argentina, Austria, Belgium, Canada, Denmark, Egypt, Ethiopia, Federal Republic of Germany, Greece, Iran, Iraq, Israel, Italy, Jordan, Kampuchea, Republic of Korea (South), Kuwait, Libya, Morocco, Netherlands, Norway, Oman, Pakistan, Peru, Portugal, Saudi Arabia, Spain, Switzerland, Taiwan, Tunisia, Turkey, United Kingdom, USA, Vietnam.

Vehicle data *Crew* 6 (commander, driver, 4 gun crew), *height* 3.289 m, *width* 3.295 m, *length* 6.612 m (with gun fully forward), *weight* 23,800 kg, *max. speed* 56 kph, *range* 360 km, *armament* 1 × 155 mm howitzer, 1 × 12.7 mm or

7.62 mm AA mg, *engine* 8V71T Detroit diesel developing 405 bhp at 2300 rpm.

Special characteristics Amphibious (using flotation bags and track propulsion at max. speed of 6.5 kph); night-driving aids; NBC protection system.

Variants M109A1 with increased effective main armament range – the barrel is longer and mounts a fume extractor (see photograph). M109A2 improved rammer and recoil system and bustle to carry additional rounds. M109G has German optical equipment. M109U of Swiss Army has semi-automatic loader. US XM975 (Roland SAM system) uses modified M109 chassis, as does the M992 Field Artillery Ammunition Support Vehicle (FAASV).

Manufacturer Allison Division of General Motors Corporation.

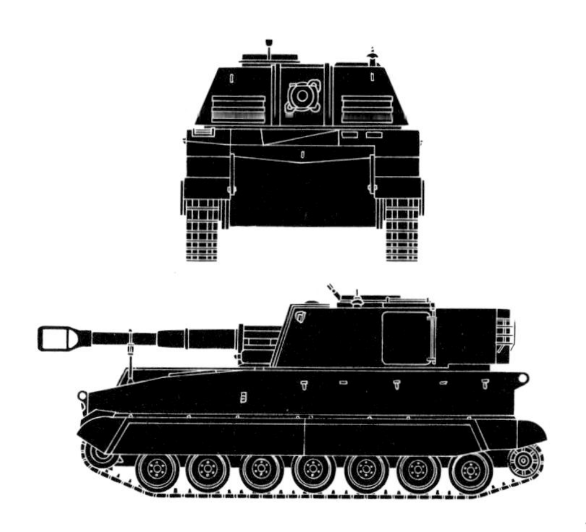

Role 203 mm (8 in.) self-propelled howitzer.

History Work started in 1956 on a new family of air-portable SP artillery weapons. These were to become M107, M110 and M578, ARV. The prototype M110 was completed in 1958. It had a petrol engine, but a year later this was altered to diesel. Production commenced in late 1961, and it entered service with the US Army in 1963, and has now replaced M107.

Employment Belgium, Federal Republic of Germany, Greece, Iran, Israel, Italy, Japan, Jordan, Republic of Korea (South), Netherlands, Pakistan, Spain, Taiwan, Turkey, United Kingdom, USA, Vietnam.

Vehicle data *Crew* 5 (commander, driver, 3 gun crew) + 8 additional gun crew on another vehicle, *height* 2.809 m, *width* 3.149 m, *length* 7.467 m (with gun fully forward), *weight* 26,550 kg, *max. speed* 56 kph, *range* 725 km, *armament* 1 × 203 mm (8 in.) howitzer, *engine* 8V71T Detroit diesel developing 405 bhp at 2300 rpm.

Special characteristics Night-driving aids, can fire nuclear rounds.

Variants M110A1 with significantly longer barrel and greater range. M110A2 with double baffle muzzle brake.

Manufacturer FMC Corporation; Pacific Car and Foundry Company; Bowen–McLaughlin–York.

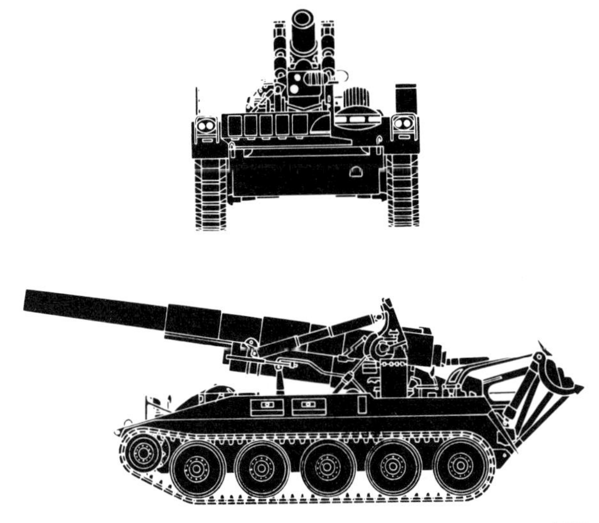

Role Armoured personnel carrier.

History In 1954 the US Army drew up a requirement for a family of airportable tracked vehicles to replace the existing M59 and M75 APCs. Prototypes were produced in 1958, and production of M113, with a petrol engine, began in 1959. In the same year work was begun to substitute a diesel engine, and production of this version, known as M113A1, started in 1963.

Employment Argentina, Australia, Belgium, Bolivia, Brazil, Canada, Chile, China (Taiwan), Costa Rica, Denmark, Ecuador, Ethiopia, Federal Republic of Germany, Guatemala, Greece (built in Italy), Haiti, Iran, Israel, Italy (built in Italy), Jordan, Kampuchea, Kenya, Republic of Korea (South), Kuwait, Laos, Lebanon, Libya (built in Italy), Morocco, Netherlands, New Zealand, Norway, Pakistan, Peru, Philippines, Portugal, Saudi Arabia, Singapore, Somalia, Spain, Sudan, Switzerland, Thailand, Tunisia, Turkey, USA, Uruguay, Vietnam.

Vehicle data *Crew* (commander, driver) + 11 infantrymen, *height* 2.5 m, *width* 2.54 m, *length* 2.686 m, *weight* 11,160 kg, *max. speed* 68 kph, *range* 480 km, *armament* 1 × 12.7 mm mg, *engine* GMC diesel 6V53 developing 215 bhp at 2800 rpm.

Special characteristics Amphibious; night-driving aids.

Variants M113A2 (automotive improvements); M113A3 (uprated engine, improved armour). Australian light reconnaissance vehicle (turret incorporating 1 × 12.7 mm and 1 × 7.62 mm mgs) and fire support vehicles (mounting Saladin, and more recently Scorpion turrets); ATGW (TOW, HOT or SS-11); M579 (fitter's vehicle with crane); mortar carriers with 120 mm Tampella (German), 107 mm (M106A2), 81 mm (M125A2) mortars; radar variants; M741 (with 6-barrelled 20 mm Vulcan AA system); M730 (with Chaparral SAM); M132A1 (flamethrower); M577A1 (command with raised roof); M548 (load carrier). Tracked Rapier on M548. M667 (Lance SSM); M727 (Hawk SAM); M981 (fire support team vehicle – FIST-V) and M113A2 chassis with Oerlikon ADATS.

Manufacturer FMC Corporation; OTO-Melara, Italy.

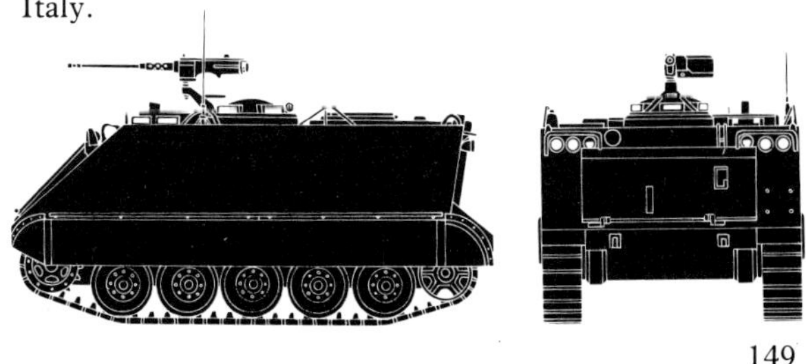

Role Mechanized infantry combat vehicle.

History This was developed privately by FMC from a modified M113A1 with turret, which was ordered and tested by US Army in the mid 1960s, but not taken up. The prototype was built in 1970, and a production order made for the Netherlands in 1975. FMC are producing M2/3 MICV for the US Army instead.

Employment Belgium, Netherlands (YPR-765 PRI), Philippines.

Vehicle data *Crew* 3 (commander, driver, gunner) + 7 infantrymen, *height* 2.784 m, *width* 2.819 m, *length* 5.258 m, *weight* 13,685 kg, *max. speed* 61 kph, *range* 490 km, *armament* 1 × 25 mm cannon, 1 × 7.62 mm coaxial mg, *engine* Detroit diesel 6V53T V-6 developing 264 bhp at 2800 rpm.

Special characteristics Amphibious (propelled by its tracks at max. speed of 6.3 kph), night-driving and -fighting aids.

Variants The Dutch have a number of variants in service – ACV; surveillance (with ZB298 radar); ambulance; 120 mm mortar tractor; cargo version; ATGW vehicle (TOW); ARV; artillery OP; AA vehicle (12.7 mm mg).

Manufacturer FMC Corporation.

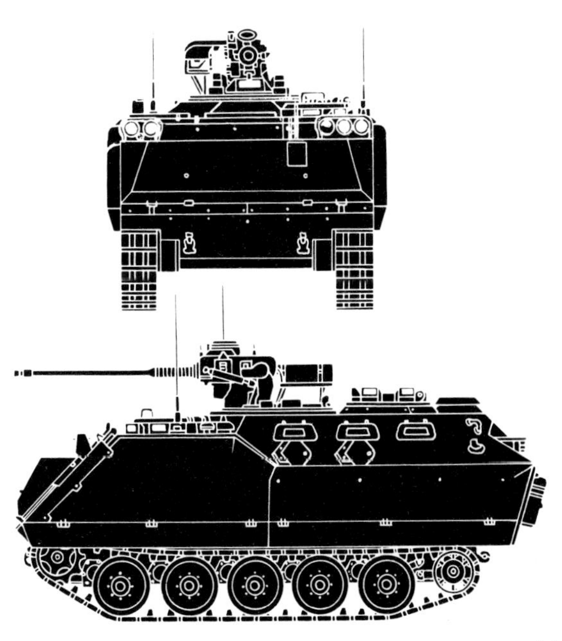

Role Mechanized infantry combat vehicle.

History Development of an MICV to replace M113 began in the mid 1960s, but early concepts were found to be too heavy and cumbersome. Then, in 1972, FMC's XM723 was selected as the basis for further work. However, at this time the US Army were looking for a suitable reconnaissance vehicle to replace the unsatisfactory M114, and two parallel projects – XM2 (infantry fighting vehicle) and XM3 (cavalry fighting vehicle) – were pursued with the main requirement that they should have the same battlefield mobility as the M1 Abrams MBT. Both entered service in 1981.

Employment USA.

Vehicle data *Crew* 3 (commander, gunner, driver) + 7 infantrymen, *height* 2.97 m, *width* 3.20 m, *length* 6.45 m, *weight* 22,680 kg, *max. speed* 66 kph, *range* 480 km, *armament* 1 × 25 mm cannon, twin TOW ATGW missile launcher, 1 × 7.62 mm coaxial mg, *engine* Cummins VTA-903 diesel developing 500 bhp at 2400 rpm.

Special characteristics Fully amphibious, being propelled by its tracks with max. speed of 6.3 kph, night-viewing and -fighting aids, 6 weapons ports in hull sides and rear using specially adapted 5.56 mm M16A1 rifle.

Variants M3 Cavalry Fighting Vehicle (CFV) has a 5-man crew (2 observers instead of infantry squad), additional ammunition, and no firing ports in hull. Fighting Vehicles Systems (FVS) Carrier is an adaptation of the M2 chassis and is to be used as a Multiple Launch Rocket System (MLRS) platform. M2A1/M3A1 are able to fire TOW 2 and have collective NBC protection.

Manufacturer FMC Corporation.

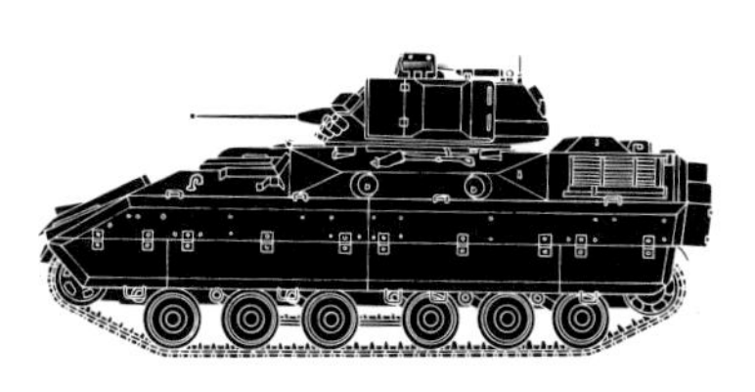

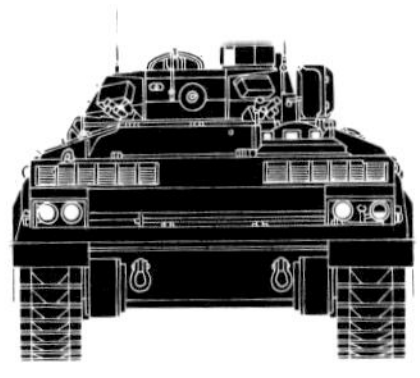

Role Multi-role light armoured vehicle.

History Design work began in 1962, the prototype appearing in 1963, and production commencing the following year. Although this was a strictly commercial undertaking, and not a result of an official US requirement, the US Army did purchase some for use by military police units in Vietnam. The initial production model was known as V-100 (see photograph). This was followed in 1969 by V-200, and in 1971 by V-150 (see silhouettes).

Employment Bolivia, Botswana, Cameroon, Dominican Republic, Ethiopia, Gabon, Guatemala, Haiti, Indonesia, Jamaica, Kuwait, Malaysia, Oman, Panama, Philippines, Saudi Arabia, Singapore, Somalia, Sudan, Taiwan, Thailand, Tunisia, Turkey, USA, Venezuela, Vietnam.

Vehicle data (V-150). *Crew* up to 12, depending on role, *height* 2.54 m (1.955 m for turretless versions), *width* 2.26 m, *length* 5.689 m, *weight* 9550 kg (with turret), *max. speed* 88 kph, *range* 700 km average, *armament* variable, *engine* Chrysler 361 V-8 petrol developing 200 bhp or Cummins V-6 diesel developing 155 bhp.

Special characteristics Amphibious (propelled by its wheels at max. speed of 4.8 kph).

Variants V-100 is lighter and slightly lower, while V-200 is significantly larger, including wheels. All models are available in a number of forms, ranging from the conventional armoured car through ATGW, APC and command versions to recovery types. Armament varies from 120 mm mortar through 90 mm gun to mgs. A new development is the V-300 6 × 6 multi-mission vehicle, with the same variety of weapons fits. An ARV version is also available.

Manufacturer Cadillac Gage Company.

Role Amphibious landing vehicle tracked personnel.

History Designed as a successor to the US Marine Corps' LVTP-5A1, development began in 1966, with the first prototype being produced the following year. Full production began in 1971 and it entered service with the US Marine Corps in 1972.

Employment Argentina, Brazil, Italy, Philippines, Republic of Korea (South), Spain, Thailand, USA, Venezuela.

Vehicle data *Crew* 3 (commander, driver, gunner) +25 infantrymen, *height* 3.263 m, *width* 3.27 m, *length* 7.943 m, *weight* 22,838 kg, *max. speed* 64 kph, *range*

480 km, *engine* Detroit diesel 8V-53T developing 400 bhp at 2800 rpm.

Special characteristics Fully amphibious with max. speed through water of 13.5 kph (water jets) and 7.2 kph (track propulsion), night-driving aids.

Variants Improved LVTP-7A1 with Cummins VT400 engine, smoke generating capability, night firing capability and other automotive improvements, LVTC-7 ACV, LVTR-7 ARV. A variant of the LVTP-7A1 mounts the Giant Viper mine clearance device.

Manufacturer FMC Corporation.

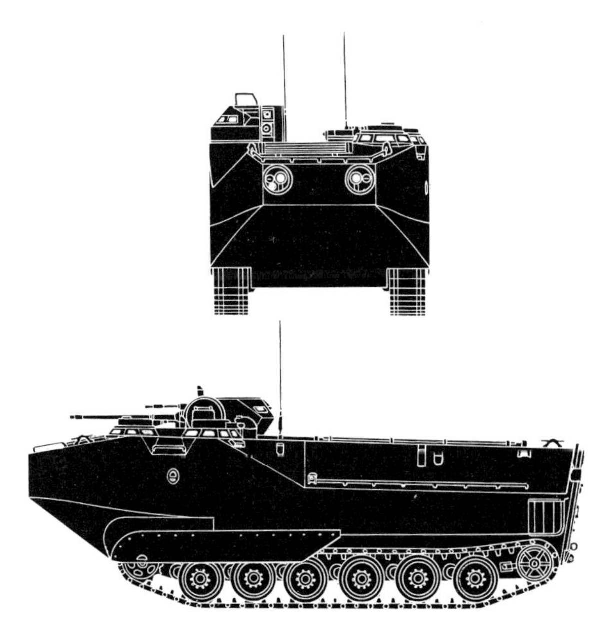

Role Main battle tank.

History T54 first appeared in 1947 as a descendant of T34 and T44. In 1961 T55 with more power and a modified turret was introduced. It is no longer in production in Warsaw Pact Countries, but is still produced in China as T59.

Employment Albania, Algeria, Afghanistan, Bangladesh, Bulgaria, China, Congo, Cuba, Cyprus, Czechoslovakia, Egypt, Ethiopia, Finland, German Democratic Republic, Guinea, Hungary, India, Iran (T59), Iraq, Israel, DPR Korea (North), Lebanon, Libya, Mongolia, Morocco, Mozambique, Nigeria, Pakistan, Palestine Liberation Army, Peru, Poland, Romania, Somalia, Sudan, Syria, Uganda, USSR, Vietnam, Yugoslavia, Zimbabwe.

Vehicle data *Crew* 4 (commander, gunner, loader, driver), *height* 2.4 m, *width* 3.27 m, *length* 6.45 m (9 m with gun fully forward), *weight* 36,000 kg, *max. speed* 48 kph (50 kph for T55), *range* 400 km (500 km for T55), *armament* 1 × 100 mm rifled gun (APHE, HEAT), 1 × 7.62 mm coaxial mg, 1 × 7.62 mm hull-mounted mg,

1 × 12.7 mm AA mg (not on earlier T55s), *engine* V-54 (V-55) diesel developing 520 bhp (580 bph for T55) at 2000 rpm.

Special characteristics Snorkel for deep fording (5.486 m maximum depth); NBC protection system on later models; night-driving and -fighting aids. Dozer blades and mine-clearing rollers can be fitted.

Variants T55 does not have a cupola for the loader, but has a turret basket. Some Indian models mount British 105 mm or Russian 115 mm gun. The Israelis have rebuilt captured Egyptian models as T55-S with British 105 mm gun and other improvements as have the British for Egypt and on T-59. The Chinese have T-69-1 (100 mm gun) and T-69-11 (105 mm rifled gun), both with laser range-finders. Various bridgelayer and ARV versions exist as well as To-55 flamethrower tank.

Manufacturer Chinese, Czech, Polish and Soviet State Arsenals.

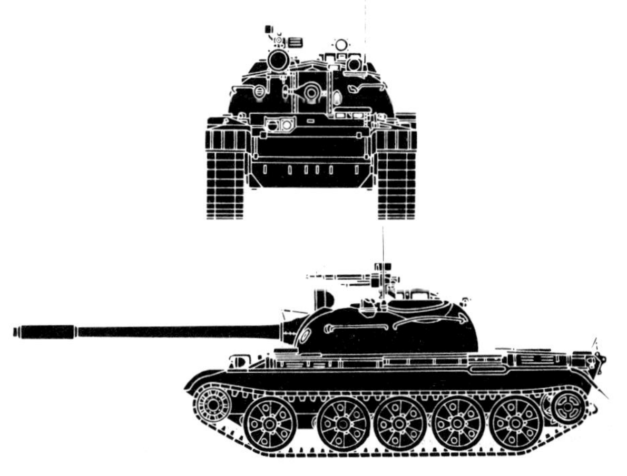

Role Light amphibious tank.

History Production commenced in 1952 and is now complete. It has been largely replaced in the Soviet Army by BMP.

Employment Afghanistan, Angola, Bulgaria, People's Republic of China, Congo, Cuba, Czechoslovakia, Egypt, Finland, German Democratic Republic, Guinea, Hungary, India, Indonesia, Iraq, Israel, DPR Korea (North), Laos, Mozambique, Pakistan, Poland, Syria, Uganda, USSR, Vietnam, Yugoslavia.

Vehicle data *Crew* 3 (commander, gunner, driver), *height* 2.2 m, *width* 3.14 m, *length* 6.91 m (7.625 m with gun fully forward), *weight* 14,000 kg, *max. speed* 44 kph, *range* 260 km, *armament* 1 × 76.2 mm rifled gun (APHE,

HEAT) 1 × 7.62 mm coaxial mg, *engine* type V-6 diesel developing 240 bhp at 1800 rpm.

Special characteristics Fully amphibious with max. speed of 10 kph.

Variants Model 1 with long multi-slotted muzzle brake; Model 2 with bore evacuator and double-baffle muzzle brake (see silhouettes); Model 3 has Model 2 muzzle brake, but no evacuator and 12.7 mm AA mg. Chinese version (known as Type 62 light tank) mounts 85 mm gun in a scaled-down T59 turret but this is now being replaced by the updated T63. Chassis is used for FROG missile launcher platform.

Manufacturer Volgograd Tank Plant.

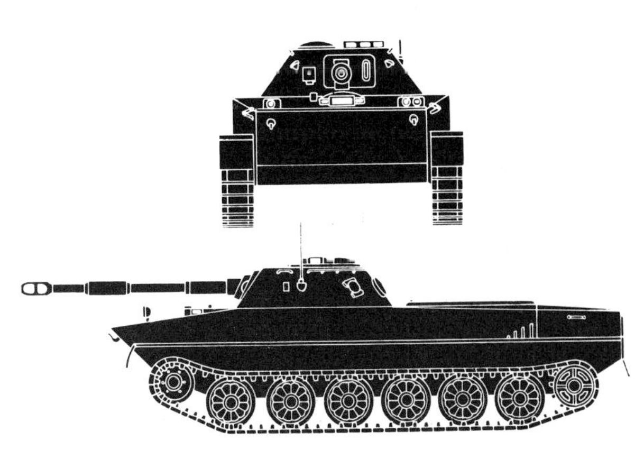

Role Main battle tank.

History Entered production in 1961–2 as a successor to T55/54. It is now being replaced in the Soviet Army by updated versions of T-64 and T-72.

Employment Afghanistan, Algeria, Angola, Bulgaria, Cuba, Czechoslovakia, Egypt, Ethiopia, German Democratic Republic, Hungary, India, Iraq, Israel, DPR Korea (North), Libya, Mongolia, Mozambique, Poland, Romania, Somalia, Syria, USSR, Vietnam, Yemen, Yugoslavia.

Vehicle data *Crew* 4 (commander, gunner, loader, driver), *height* 2.4 m, *width* 3.35 m, *length* 6.715 m (9.77 m with gun fully forward), *weight* 36,000 kg, *max. speed* 50 kph, *range* 500 km, *armament* 1 × 115 mm smoothbore gun (APDSFS, HEAT, HE), 1 × 7.62 mm coaxial mg, *engine* V-2-62 V-12 diesel developing 700 bhp at 2200 rpm.

Special characteristics NBC protection system; night-fighting and -driving aids; deep-fording capability using snorkel.

Variants T62A has 12.7 mm AA mg on loader's hatch (see photograph). T-62K command version with additional aerial bases. T-62M is T-62A with T-72 track and drive sprocket. Flamethrower and ARV versions also exist, and recently smoke grenade dischargers have been mounted on the turret. Britain also offers a rifled 115 mm gun for mounting in T-62.

Manufacturer Soviet State Arsenals.

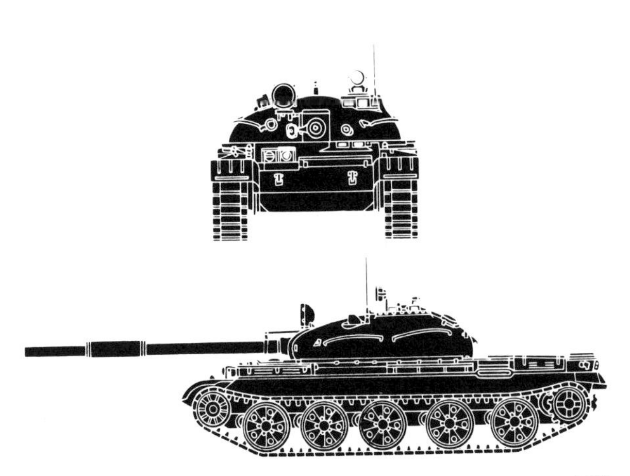

Role Main battle tank.

History This evolved from the experimental T-70, which was armed with the same 115 mm as T-62, and entered service in the late 1960s. T-72 was developed in parallel to T-64. T-64 has, in the past, had automotive problems, which probably explains why it has not appeared outside USSR. Production is now completed.

Employment USSR.

Vehicle data *Crew* 3 (commander, driver, gunner), *height* 2.265 m (excluding AA mg), *width* 3.375 m, *length* 9.02 m (gun fully forward), *weight* 38,000 kg, *max. speed* 70 kph, *range* 500 km, *armament* 1 × 125 mm smoothbore (APDSFS, HEAT), 1 × 7.62 mm coaxial mg, 1 × 12.7 mm AA mg, *engine* diesel developing 700 bhp at 2200 rpm.

Special characteristics Night-driving and -fighting aids; NBC protection system; ability to snorkel; autoloader and optical rangefinder. Later models are likely to have laser rangefinder and Chobham-type armour.

Variants T-64K command tank has no 12.7 mm AA mg, and sports a large antenna when static. T-64A (M1980) with turret mounted smoke grenade dischargers and narrow track guard covering jockey wheels; T-64B with larger track guard, enlarged optical device LH turret side, together with Kobra ATGW system.

Manufacturer Soviet State Arsenals.

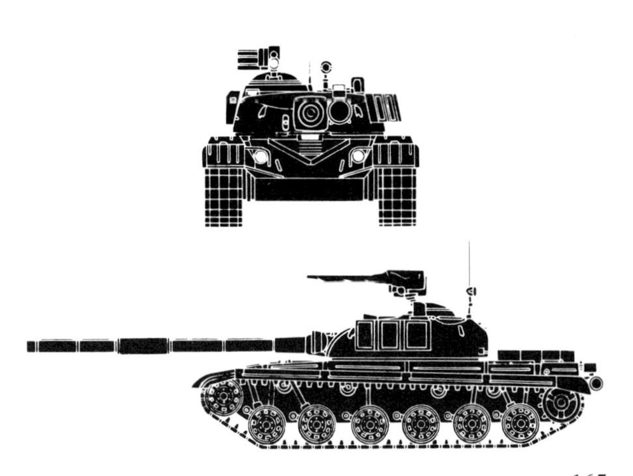

Role Main battle tank.

History As with T-64, T72 owes its origins to the experimental T-70, and was developed in parallel with T-64, although it went into production slightly later.

Employment Algeria, Bulgaria, Cuba, Czechoslovakia, German Democratic Republic, Hungary, India, Iraq, Libya, Poland, Romania, Syria, USSR, Yugoslavia.

Vehicle data *Crew* 3 (commander, driver, gunner), *height* 2.265 m, (excluding AA mg), *width* 3.375 m, *length* 9.02 m (gun fully forward), *weight* 40,000 kg, *max. speed* 70 kph, *range* 500 km, *armament* 1 × 125 mm smoothbore gun (APDSFS, HEAT), 1 × 7.62 mm coaxial mg, 1 × 12.7 mm AA mg, *engine* diesel developing 700 bhp at 2200 rpm.

Special characteristics Night-fighting and -driving aids. NBC protection system, ability to snorkel. Autoloader, optical rangefinder.

Variants Later versions: M1980 (T-74) with no optical rangefinder visible (possibly updated by laser rangefinder); M1981 with banks of smoke-grenade dischargers on turret front (see photograph). Both are likely to have Chobham-type armour and US sources have designated them T-80. Recent models also have the Kobra ATGW system. A command version exists without AA mg, but including additional radio antennae, and radio mast mounted on turret when static. It is possible that export models may be of a lower standard than those in Warsaw Pact service. BREM-1 ARRV on T-72 chassis also exists.

Manufacturer Soviet State Arsenals.

Role Reconnaissance vehicle.

History BRDM-1 (also known as BTR-40P) with an open turret was first noted in 1959 as a replacement for BTR-40. The improved BRDM-2 (BTR-40PV) with enclosed turret entered service with the Soviet Army in the early 1960s.

Employment Albania, Algeria, Angola, Benin, Bulgaria, Cape Verde Islands, Central African Republic, Chad, Congo, Cuba, Egypt, Equatorial Guinea, Ethiopia, German Democratic Republic, Guinea, Hungary, Iraq, Israel, Jibuti, Libya, Madagascar, Malawi, Mali, Mauritania, Mongolia, Morocco, Mozambique, Nicaragua, Peru, Poland, Romania, Sao-Tome Principe, Seychelles, Somalia, Sudan, Syria, Tanzania, Uganda, USSR, Vietnam, Yemen, Yugoslavia, Zambia, Zimbabwe.

Vehicle data *Crew* 4 (commander, gunner, two drivers), *height* 2.31 m, *width* 2.35 m, *length* 5.75 m, *weight* 7,000 kg, *max. speed* 100 kph, *range* 750 km, *armament* 1 × 14.5 mm, 1 × 7.62 mm coaxial mg, *engine* GAZ-41, V8 petrol developing 140 bhp at 3400 rpm.

Special characteristics Amphibious, being propelled by a single water jet at rear of hull with max. speed of 10 kph; night-fighting and -driving aids; NBC protection system; 2 × 2 retractable cross-country wheels.

Variants BRDM-1 with open turret, which is also used as NBC reconnaissance vehicle and ACV, as well as in ATGW role using Snapper (3 missile launchers), Swatter (4 missile launchers), Sagger (6 missile launchers); BRDM-2 ACV with turret removed; NBC reconnaissance; Sagger ATGW (6 missile launchers), Swatter (4 launchers), Spandrel ATGW (5 launchers), SAM-7 (8 launchers); SAM-9 (4 launchers).

Manufacturer Soviet State Arsenals.

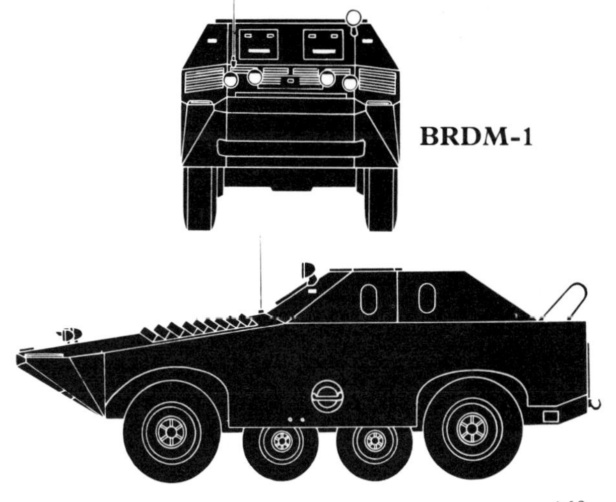

BRDM-1

Role 152 mm self-propelled howitzer.

History This represented a radical new departure for the Soviet Union, since up until the early 1970s she had relied almost entirely on towed artillery. It was developed during the late 1960s, and first seen in public in 1974.

Employment German Democratic Republic, Iraq, Libya, USSR.

Vehicle data *Crew* 3 (commander, driver, gunner) + additional crew carried in separate vehicle, *height* 2.7 m, *width* 3.2 m, *length* 7.7 m, *weight* 27,000 kg, *max. speed* 70 kph, *range* 400 km approx., *armament* 1 × 152 mm, *engine* V-12 diesel developing 520 bhp.

Special characteristics Night-driving aids; NBC protection system.

Variants Chassis is used for GANEF SAM-4 launcher vehicle, M-1975 240 mm SP mortar, 2S5 152 mm SP gun, M-1975 203 mm SP gun.

Manufacturer Soviet State Arsenals.

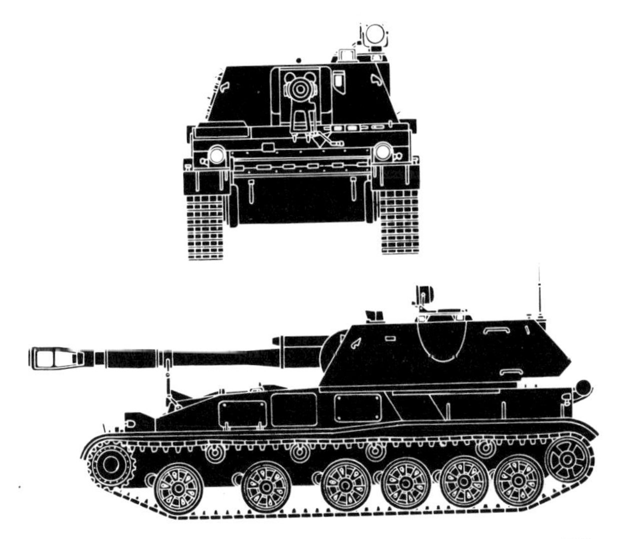

Role Self-propelled howitzer.

History M-1974 made its appearance at the same time as M-1973. It uses a chassis similar to PT-76, but with an additional pair of road wheels.

Employment Algeria, Angola, Czechoslovakia, Ethiopia, German Democratic Republic, Hungary, Iraq, Libya, Poland, Syria, USSR, Yugoslavia.

Vehicle data *Crew* 3 (commander, driver, gunner) + additional gun crew carried in separate vehicle, *height* 2.37 m, *width* 2.95 m, *length* 7.1 m, *weight* 20,000 kg, *max. speed* 70 kph, *range* 400 km (approx.), *armament* 1 × 122 mm, *engine* diesel, but no other details known.

Special characteristics Night-driving aids; NBC protection system. It is also amphibious with max. water speed of 5–10 kph. It has an anti-armour capability, using HEATFS, and a rocket-assisted projectile has also recently been introduced.

Variants ACRV-2 command and reconnaissance vehicle with each 6-gun battery; mine-clearing vehicle with rocket-propelled mine-clearing system. IPR amphibious vehicle and RkhM chemical reconnaissance vehicle use the same chassis.

Manufacturer Soviet State Arsenals.

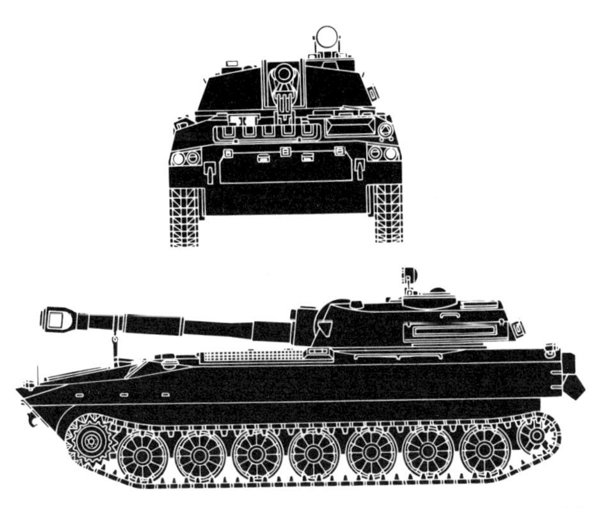

Role Armoured personnel carrier.

History Based on the ZIL-151 and -157 truck chassis, this was the first Soviet-built APC, and entered service in the late 1940s, and is still in service with Warsaw Pact second-line units.

Employment Afghanistan, Albania, Algeria, Angola, Bulgaria, Central African Republic, Chad, People's Republic of China (as Type 56 APC), Congo, Cuba, Cyprus, Egypt, Ethiopia, German Democratic Republic, Guinea, Guinea-Bissau, Hungary, India, Indonesia, Iran, Iraq, Israel, Kampuchea, DPR Korea (North), Laos, Mali, Mongolia, Mozambique, Nicaragua, Poland, Romania, Somalia, Sri Lanka, Sudan, Syria, Tanzania, Uganda, USSR, Vietnam, Yemen, Yugoslavia, Zaire, Zimbabwe.

Vehicle data *Crew* 2 (commander, driver) + 17 infantrymen, *height* 2.05 m, *width* 2.32 m, *length* 6.83 m, *weight* 8950 kg, *max. speed* 75 kph, *range* 650 km, *armament*

1 × 7.62 mm mg, *engine* ZIL-123 petrol developing 110 bhp at 2900 rpm.

Special characteristics Nil.

Variants Earlier versions had an open top. BTR-152U ACV (with higher roof); BTR-152A AA vehicle (2 × 14.5 mm mgs). Some Soviet models have ATGW Sagger launchers, and some Egyptian models have 4 × 12.7 mm Czech mgs.

Manufacturer Soviet State Arsenals.

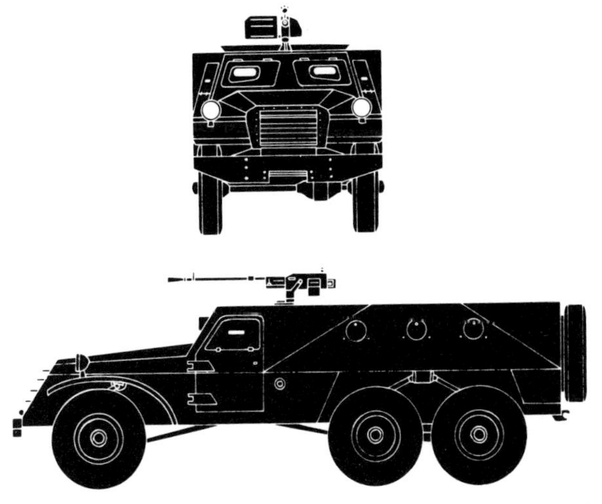

Role Armoured personnel carrier.

History BTR-50 was first seen in public in 1957. Although now found only in Soviet second-line units, it is still widely used by many countries throughout the world. It uses the PT-76 chassis.

Employment Afghanistan, Albania, Algeria, Angola, Bulgaria, Congo, Cyprus, Egypt, Finland, German Democratic Republic, Guinea, India, Iran, Iraq, Israel, DPR Korea (North), Libya, Romania, Somalia, Sudan, Syria, USSR, Vietnam, Yugoslavia.

Vehicle data *Crew* 2 (commander, driver) + 20 infantrymen, *height* 1.97 m, *width* 3.14 m, *length* 7.08 m, *weight* 14,200 kg, *max. speed* 44 kph, *range* 260 km, *armament* 1 × 7.62 mm mg, *engine* type V-6 diesel developing 240 bhp at 1800 rpm.

Special characteristics Amphibious (propelled by two water jets at rear of hull with max. speed of 10 kph); night-fighting and -driving aids.

Variants BTR-50P (earliest model with open crew

compartment); BTR-50PK (closed crew compartment and NBC protection system); BTR-50PU ACV. BTR50MTP ARV; BTR50MTK mine-clearing vehicle; ambulance; mortar carrier. Czech OT-62 is based on this vehicle (see separate entry). China has a similar APC called Type 77.

Manufacturer Soviet State Arsenals.

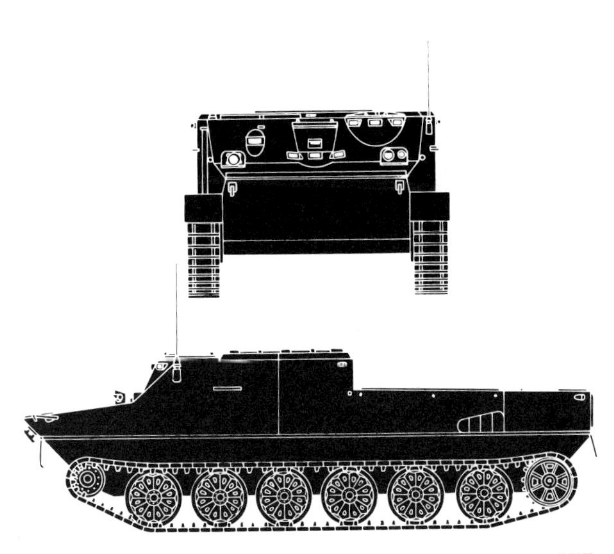

Role Armoured personnel carrier.

History The successor to BTR-152, BTR-60 entered production in 1960 and was first seen a year later.

Employment Afghanistan, Algeria, Angola, Botswana, Bulgaria, Chad, China, Congo, Cuba, Egypt, Ethiopia, Finland, German Democratic Republic, Guinea, Hungary, Iran, Iraq, Israel, Jibouti, Kampuchea, DPR Korea (North), Libya, Mali, Mongolia, Mozambique, Nicaragua, Romania, Somalia, Syria, USSR, Vietnam, Yemen, Yugoslavia, Zambia.

Vehicle data *Crew* 2 (commander, driver) + 14 infantrymen, *height* 2.31 m, *width* 2.825 m, *length* 7.56 m, *weight* 10,300 kg, *max. speed* 80 kph, *range* 500 km, *armament* 1 × 14.5 mm mg, 1 × 7.62 mm coaxial mg, *engine* 2 × GAZ-49B petrol developing 90 bhp each at 3400 rpm.

Special characteristics Amphibious (propelled by a single hydro-jet at max. speed of 10 kph); NBC protection system; night-fighting and -driving aids.

Variants BTR-60P (original model with open-topped crew compartment and no turret); BTR-60PU ACV (canvas roof); BTR-60PA (or K) (armoured protection over crew compartment, but no turret); BTR-60PB FAC vehicle (plexiglass window replaces turret armament, and with portable generator on rear deck); BTR-60MS radio vehicle with telescopic antennae. BTR-70 is a later development with lengthened hull (8.0 m) and larger engine compartment. Variants include version with AGS-17 automatic grenade launcher, BTR-70MS radio vehicle (turretted), BTR-70KShM ACV and BREM ARRV. Romanian version (TAB-72) has different turret and TAB-73 has no turret, but 82 mm mortar. TAB-77 is the Romanian version of BTR-70.

Manufacturer Soviet State Arsenals.

BMP-1

Role Mechanized infantry combat vehicle.

History First seen in 1967, it was the first MICV to enter service worldwide. It uses many components common to PT-76.

Employment Afghanistan, Algeria, Cuba, Czechoslovakia, Egypt, Ethiopia, Finland, German Democratic Republic, Hungary, India, Iran, Iraq, DPR Korea (North), Libya, Mongolia, Poland, Syria, USSR, Yugoslavia.

Vehicle data *Crew* 3 (commander, gunner, driver) + 8 infantrymen, *height* 1.98 m, *width* 2.97 m, *length* 6.75 m, *weight* 12,500 kg, *max. speed* 55 kph, *range* 300 km, *armament* 1 × 73 mm smooth-bore gun, 1 × Sagger ATGW launcher, 1 × 7.62 mm coaxial mg, *engine* type V-6 diesel developing 280 bhp at 2000 rpm.

Special characteristics Amphibious (propelled by its tracks at max. speed of 8 kph); NBC protection system; night-driving and -fighting aids.

Variants Surveillance model with radar mounted in place of crew compartment exists, and also a reconnaissance version without Sagger, command and communications versions. BMP-2 (M1981) was first seen in 1982. It has a 2-man turret armed with a 30 mm cannon and a Spandrel ATGW launcher. Turret is further to rear than BMP-1 and BMP-2 carries only 6 infantrymen, their compartment having two rather than four roof hatches.

Manufacturer Soviet State Arsenals.

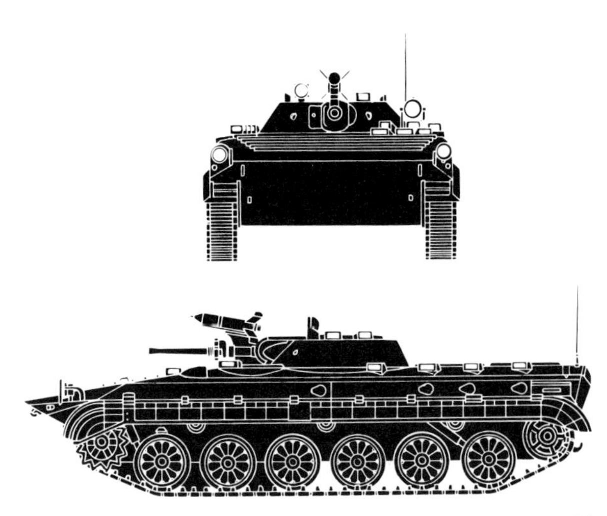

Role Airborne infantry combat vehicle.

History BMD-1 was first seen on public display in late 1973. It has the same turret as BMP-1, and was probably designed to replace both ASU-57 and ASU-85.

Employment India, USSR.

Vehicle data *Crew* 3 (commander, gunner, driver) + 6 infantrymen, *height* 1.85 m, *width* 2.65 m, *length* 5.3 m, *weight* 9000 kg, *max. speed* 80 kph, *range* 300 km (approx.), *armament* 1 × 73 mm smooth-bore gun, 1 × Sagger ATGW launcher, 3 × 7.62 mm mgs (1 coaxial, 2 hull-mounted), *engine* V-6 liquid-cooled diesel developing 290 bhp.

Special characteristics Amphibious (propelled by water jets at rear of hull with max. speed of 6 kph); night-fighting and -driving aids; NBC protection system. Adjustable suspension, giving variable ground clearance.

Variants BMD-1M with 30 mm cannon; version with Spigot ATGW and 82 mm mortar version. BMD-2 has an additional pair of road wheels and the command version of this has no turret.

Manufacturer Soviet State Arsenals.

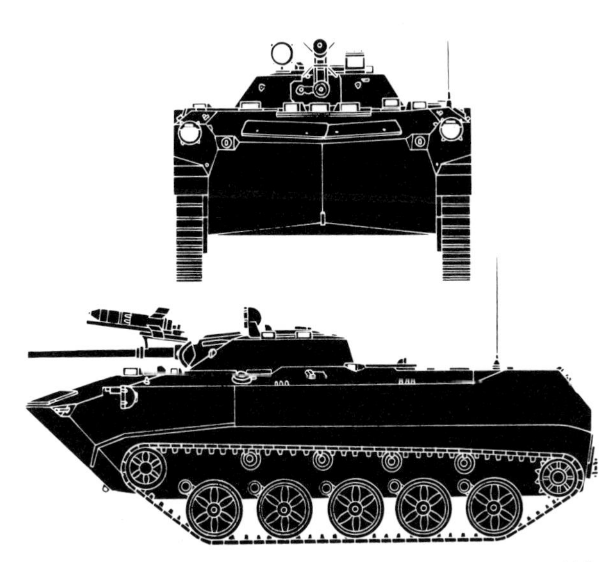

Role Self-propelled anti-aircraft gun system.

History First seen in public in 1965, ZSU 23-4 uses the same chassis as PT-76, and is still widely seen in Warsaw Pact front-line formations.

Employment Afghanistan, Angola, Algeria, Bulgaria, Cuba, Czechoslovakia, Egypt, Ethiopia, Finland, German Democratic Republic, Hungary, India, Iran, Iraq, Jordan, DPR Korea (North), Libya, Mozambique, Nigeria, Poland, Peru, Romania, Somalia, Syria, USSR, Vietnam, Yemen, Yugoslavia.

Vehicle data *Crew* 4 (commander, driver, gunner radar operator), *height* 2.25 m (radar retracted), *width* 2.95 m, *length* 6.3 m, *weight* 14,000 kg, *max. speed* 44 kph, *range* 260 km, *armament* 4 × 23 mm cannon, *engine* type V-6 diesel developing 240 bhp at 1800 rpm.

Special characteristics Night-driving aids; NBC protection system. The guns are radar-controlled, but it also has optical sights. Effective range is 3000 m.

Variants Model A had two small bins on each side of turret, while Model B (as shown) has long bins on either side. Other versions with minor differences exist.

Manufacturer Soviet State Arsenals.

Role Airportable tank destroyer.

History ASU-85 made its first public appearance in 1962. It was designed specifically to provide integral anti-tank capability to airborne forces. It uses many PT-76 components.

Employment German Democratic Republic, Poland, USSR.

Vehicle data *Crew* 4 (commander, gunner, loader, driver), *height* 2.1 m, *width* 2.8 m, *length* 8.49 m (including gun), *weight* 14,000 kg, *max. speed* 45 kph, *range* 260 km, *armament* 1 × 85 mm rifled gun (HE, APHE, HVAP), 1 × 7.62 mm coaxial mg, *engine* type V-6 diesel developing 240 bhp at 1800 rpm.

Special characteristics Night-fighting and -driving aids; NBC protection system.

Variants Since 1970 ASU-85 has been fitted with 12.7 mm AA mg.

Manufacturer Soviet State Arsenals.

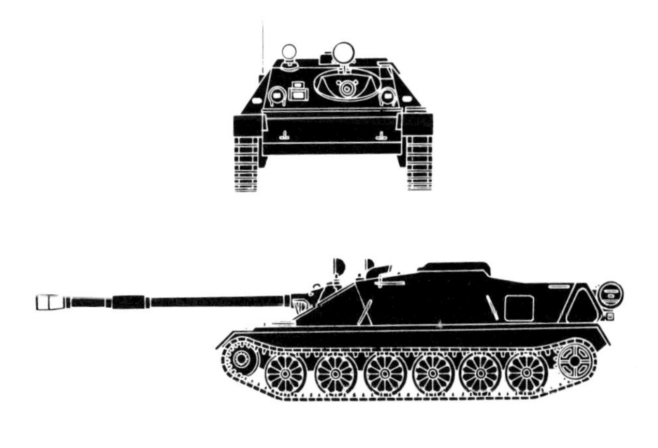

LIST OF ADDRESSES

Austria
Steyr-Daimler-Puch AG, Werke Wien, A-1111 Vienna

Brazil
Engesa SA, avenida Načoes Unidas 22.833, CEP 04795, POB 6637, São Paulo

Canada
Diesel Division, General Motors Canada, PO Box 5160, London, Ontario N6A 4N5

France
ATS Roanne, Groupement Industriel des Armaments Terrestres, 10 place Georges Clémenceau, 92211 St Cloud
Creusot-Loire Division de la Mécanique Spécialisée, 15 rue Pasquier, 75383 Paris, Cedex 08
Saviem Creusot-Loire, 316 Bureaux de la Colline, 92213 Saint-Cloud
Société Automobiles M. Berliet, 160 boulevard de Verdun, 92400 Courbevoie
Société de Construction Mécanique Panhard et Levassor, 18 avenue d'Ivry, 75013 Paris

Federal Republic of Germany
Klockner-Humboldt-Deutz AG, Deutz-Mülheimer-Strasse 111, D-5000 Köln-Deutz
Krauss-Maffei AG, Ordnance Division, Krauss-Maffei-Strasse 2, D-8000 Munich 50
Dr Ing. LCF Porsche Atkiengesellschaft D-7000 Stuttgart 40
Rheinstahl Wehrtechnik, Kassel
Thyssen-Henschel, Postfach 102969, D-3500 Kassel

Israel
Ramta Structures and Systems, POB 323, Beersheba
Soltam Ltd, POB 371, Haifa

Italy
OTO-Melara, Via Valdilocchi 15, 19100 La Spezia

Japan
Mitsubishi Heavy Industries, 5-1 Marunouchi 2-chrome, Chiyoda-Ku, Tokyo

Netherlands
DAF, Van Doorne's Automobielfabrieken, Geldropseweg 303, Eindhoven

Portugal
BRAVIA SARL, Av. Eng. Duarte Pachico 21, 5°-A Lisbon

Spain
Empresa Nacional de Autocamiones SA, Avda. de Aragón 402, 28022 Madrid

South Africa
Sandoek-Austral Naperk Ltd, PO Box 6390, West Street Industrial Sites, Boksburg, Transvaal

Sweden
AB Bolors, Ordnance Division, Box 500, S-690 20 Bofors
AB Hägglunds and Soner, Örnsköldsvik

Switzerland
Federal Construction Works, Thun
MOWAG Motorwagenfabrik AG, 8280 Kreuzlingen

United Kingdom

Alvis Ltd, Holyhead Road, Coventry

GKN Sankey Ltd, Telford, Salop

Marketing Coordinator, Royal Ordnance factories, Ministry of Defence, St Christopher House, Southwark St, London SE1 0TD

Short Bros. & Harland Ltd, PO Box 241, Airport Road, Belfast BT3 902, Northern Ireland

Vickers Ltd, Elswick, Newcastle-upon-Tyne

USA

Allison Division of General Motors Corporation, Cleveland, Ohio

Cadillac Gage Company, PO Box 1027, Warren, Michigan 48090

General Dynamics Land Systems Division, PO Box 1901, Warren, Michigan 48090

FMC Corporation, Ordnance Division, 1105 Coleman Avenue, San José, California 91508

INDEX

TANKS/TANK DESTROYERS

AMX-13 (France) *26–7, 31*
AMX-30 (France) *28–9, 52*
ASU-85 (USSR) *186–7*
Centurion (UK) *106–7*
Challenger (UK) *109, 110–11*
Chieftain (UK) *52, 108–9*
EE-T1 Osorio (Brazil) *16–17*
Gen Leclerc (France) *8*
IKV-91 (Sweden) *98–9*
Jagdpanzer SK 105 Kürassier (Austria) *14–15, 27*
Khalid (Jordan) *108–9*
Leopard 1 (FRG) *52–3, 76*
Leopard 2 (FRG) *54–5*
M1 Abrams (USA) *138–9*
M48 (USA) *7–8, 134–5*
M60 (USA) *136–7*
Merkava (Israel) *70–1*
OF-40 (Italy) *76–7, 80*
Olifant (S Africa) *106–7*
PT-76 (USSR) *160–1*
Pz 61, Pz 68 (Switzerland) *102–3*
Securi (Brazil) *27*
Shir (Iran) *108–9*
Strv 101/102/104 (Sweden) *106–7*
Strv 103 (Sweden) *6, 96–7*
T-54/55 (USSR) *8, 158–9*
T-59 (China) *158–9*
T-62 (USSR) *162–3*
T-62/63 (China) *160–1*
T-64 (USSR) *6, 7, 166–7*
T-69 (China) *158–9*
T-72/74/80 (USSR) *6, 7, 166–7*
TAM (Argentina) *56, 57, 81*
TH 301 (FRG) *56–7*
Type 74 (Japan) *82–3*
Valiant (UK) *112–13*
Vickers (UK) *112–13*
Vijayanta (India) *112–13*

RECONNAISSANCE VEHICLES

AMX-10RC (France) *38–9*
BMR-VEC (Spain) *94–5*
BRDM (USSR) *68, 168–9*
Chaimite (Portugal) *90–1*
Commando (USA) *90, 154–5*
Cougar (Canada) *105*
EE-9 Cascavel (Brazil) *18*
Eland (S Africa) *32, 92*
Ferret (UK) *114–15*
Fox (UK) *118, 120–1*
FUG-63 (Hungary) *68–9*
Lanza (Brazil) *37*
Luchs (FRG) *58–9, 62*
Lynx (Argentina) *37*
Lynx (M113 C&R) (USA) *140–1*
M3 Bradley (USA) *152–3*
M11 VBL (France) *6, 40–1*
OT-65 (Czechoslovakia) *68–9*
Panga (UK) *121*
Panhard AML 90 (France) *32–3*
Panhard EBR 75 (France) *30–1*
Panhard ERC 90 (France) *36–7*
Piranha (Switzerland) *104–5*
RBY (Israel) *72–3*
Renault VBC 90 (France) *34–5*

Saladin (UK) *116–17*
Scimitar (UK) *119*
Scorpion (UK) *116, 118–19*
Shark (Chile) *105*
Shorland (UK) *130–1*
Spy (Switzerland) *105*
Type 87 (Japan) *87*
Type 6616 (Italy) *78–9*
Weisel (FRG) *5, 6, 66–7*

APCs/MICVs

AMX-10P (France) *46–7, 48*
AMX-10RTT (France) *39*
AMX-13 VCI (France) *43, 46*
BMD (USSR) *182–3*
BMP (USSR) *180–1*
BTR-50 (USSR) *176–7*
BTR-60 (USSR) *178–9*
BTR-70 (USSR) *179*
BTR-152 (USSR) *174–5*
EBR ETT (Portugal) *31*
FMC M765 (USA) *88, 150–1*
Fuchs (FRG) *62–3*
FV432 (UK) *124–5*
Grizzly (Canada) *105*
LVTP-7 (USA) *156–7*
M2 Bradley (USA) *152–3*
M113A1 (USA) *148–9*
Marder (FRG) *56, 60–1*
OT-62B (Czechoslovakia) *22–3*
OT-64 (Czechoslovakia) *24–5*
Ratel (S Africa) *92–3*
Samaritan (UK) *132–3*
Samson (UK) *132–3*
Saracen (UK) *92*
Saviem/Creusot-Loire VAB (France) *48–9*
Saxon AT 105 (UK) *124, 126–7*
Spartan (UK) *132–3*
Steyr 4K 7FA-K SPz (Austria) *12–13*
Striker (UK) *132–3*
Sultan (UK) *132–3*
TOPAS (Poland) *22–3*
Type 73 (Japan) *84–5*
Type 82 (Japan) *86–7*
Type 6614 (S Korea) *79*
Urutu EE-11 (Brazil) *20–1*
Warrior MCV-80 (UK) *128–9*
YP-408 (Netherlands) *88–9*
YPR-765 PRI (Netherlands) *88, 150–1*

SP ARTILLERY

155 mm Mk F3 (France) *42–3*
2S1 (USSR) *172–3*
2S3 (USSR) *170–1*
Abbot (UK) *122–3*
AMX GCT (France) *44–5*
L-33 (Israel) *74–5*
M107 (USA) *43, 142–3*
M109 (USA) *144–5*
M110 (USA) *142, 143, 146–7*
Palmaria (Italy) *77, 80–1*
SP-70 (FRG/Italy/UK) *8, 122*
VK 155 Bandkanon 1A (Sweden) *100–1*

AIR DEFENCE

AMX-13 DCA (France) *50–1*
Cheetah (Netherlands) *64–5*
Gepard (FRG) *53, 64–5*
ZSU 23-4 (USSR) *184–5*